THE DRINKING WATER HAN

THE DRINKING WATER HANDBOOK

Frank R. Spellman, Ph.D.
Joanne Drinan

Hampton Roads Sanitation District

TECHNOMIC
PUBLISHING CO., INC.
LANCASTER · BASEL

The Drinking Water Handbook

aTECHNOMIC®publication

Technomic Publishing Company, Inc.
851 New Holland Avenue, Box 3535
Lancaster, Pennsylvania 17604 U.S.A.

Printed in the United States of America
10 9 8 7 6 5 4 3 2 1

Main entry under title:
    The Drinking Water Handbook

A Technomic Publishing Company book
Bibliography: p.
Includes index p. 257

Library of Congress Catalog Card No. 99-69360
ISBN No. 1-56676-833-0

To
Teresa Wiegand,
as always, it is a pleasure working with you.

And to
Susan Farmer and young Anne Victoria,
congratulations!

And to
Kimberly Martin,
congratulations!
—Frank Spellman

To
my husband, John, and
my son, Michael, for their understanding.
—Joanne Drinan

Table of Contents

Preface

NOTWITHSTANDING our absolute need to breathe untainted air, nothing is more important to us than the quality of the water we drink, though, of course, we need clean water for other uses as well.

All the major cities of the modern world grew up on waterfronts. This is not because people require such large amounts of water for survival. While people typically require no more than 10 pounds of water to create each pound of flesh, to make a pound of paper takes approximately 250 pounds of water, and 600 to produce one pound of fertilizer. Our cities grew large near water primarily because industry demands a water supply. In the United States, industry uses over 100 cubic miles of water every year to cool, wash, and circulate its material, an amount equal to 30% of all the water in the rivers of the world. Of this water we use, very little goes back cleaner than when taken from its source—because, as water travels, it bears with it the story of where it has been and what it has been used for.

This text recognizes the value of water for use in industry, but is not about the industrial use of freshwater. The focus of this text is freshwater's primary purpose, human use. We need pure, sweet clean water to sustain us. This text is about the technology available and required to ensure that the water from our taps is safe.

The Drinking Water Handbook focuses on ensuring safe drinking water, and the primary focus is on discussing current problems with our drinking water supply, and the technologies available to mitigate the problems. The discussion in this text relating to solutions and technologies is not the result of a "feel good" approach, but rather is based on science and technology.

Concern over water quality is not new. Throughout the history of human civilization, the concern over the availability of clean drinking water has played an instrumental role in determining where people chose to settle and how these settlements grew into the cities of today. Those of us who reside in the United States are blessed with an abundant freshwater supply. Technology has even allowed us to provide for our arid areas. However, even with our abundance, economic development and population growth strain the quality and quantity of water available for drinking.

Trillions of gallons of precipitation fall on the U.S. every day, filling streams, rivers, ponds, lakes, and marshes. That water then percolates through soil's natural filter to recharge our underground freshwater aquifers. And each day, agricultural irrigators, industrial users, factories, and homeowners withdraw hundreds of billions of gallons from this finite water supply. We use this water for everything from washing dishes and watering the garden to cooling the equipment of industrial complexes. After we are finished with it, the water (a substance always and forever in motion) finds a path back into the water cycle—into a stream, river, pond, lake, marsh, or groundwater supply—along with whatever contaminants it picked up along the way.

When we open our taps to fill our glasses with drinking water, we demonstrate that we expect good quality water as our right. As far as most of us are concerned, what comes from the tap is safe, a substance that will cause us no harm. Is this really the case? Is the water from our taps safe? We are hearing now that cancer-causing chemicals exist in virtually every public water supply in the United States.

As the water pours from the tap into our drinking glass, there is another point of concern. Has the water been tested in accordance with applicable standards or requirements? Were the tests reliable, or not? Most public health officials claim our drinking water is safe—do they really know whether it is safe or not? Are federal and state standards for water safety adequate, or not?

The Drinking Water Handbook provides technical information about what is in many tap water supplies, and the measures necessary to ensure the safety of consumers. *The Drinking Water Handbook* starts at the water source itself, and describes the water purification process through distribution to the tap, to our actual use and reuse of water.

Water, a substance we constantly use and reuse, is recycled via the hydrologic (water) cycle. This text will focus on a particular water cycle, the artificial water cycle we created, control, and are utterly dependent on. Called the Urban Water Cycle, it consists of the water supply, water purification, water use, and water disposal for reuse cycle common in major metropolitan areas—a man-made cycle that mimics the natural water cycle.

As water users directly affected by water's quality, one of the most important steps we can take to protect our health is to make sure that the water available for drinking is safe—a step not easy to accomplish. We can't tell the quality of our water just by looking at it; we know that water can look clear in the glass and still contain toxic chemicals or bacterial and viral pathogens that can make users sick.

To purify water, communities use municipal treatment plants and a variety of technologies ranging from simple screens, sand filtration, and disinfection, to complex chemical and mechanical processes. However, these systems are not fail-safe. When they do fail, water users are left vulnerable to a wide variety of biological and chemical hazards.

In 1993, a microscopic organism called *Cryptosporidium* caused more than 400,000 illnesses in Milwaukee, Wisconsin, and left 100 people dead. In 1994, two more outbreaks of the same protozoan killed 19 and sickened more than 100 in Las Vegas. Panic over *Cryptosporidium* and *Giardia* caused a two-month-long boil-alert crisis in Sydney, Australia between July and September of 1998, one that ended up costing millions of dollars, though no illnesses resulted. The end result: that treatment facility is paying enormous penalties for incompetent testing, and for not following the maxim, "it is better to be correct than sorry."

Many water users and technologists are no longer ignorant of the current drinking water crisis—the publicity generated by the events in Milwaukee, Las Vegas, and Sydney, Australia stories took care of that shortcoming. The infamous *Cryptosporidium* outbreak had a tornado-like effect. Microbiological parameters and controls returned to the forefront, after having been displaced in the 1970s; disinfection, along with more sophisticated water treatment, was back in favor. Overnight, *Cryptosporidium* and *Giardia* became urgent targets of concern, and fear of carcinogens (from radon, lead, and arsenic) was no longer solely at the top of the regulatory agenda. Then, in late 1998, the two new targets of concern, *Cryptosporidium* and *Giardia*, were joined by not necessarily a new concern, but a concern with new emphasis: disinfection by-products (DBP) such as halogenated chloro-organic compounds—trihalomethanes (THMs). A partial result of this widespread concern is the emergence of a new "packaged" water industry, one growing at tremendous speed, because consumers want assurances that their water is safe no matter what.

In the not too distant past, determining whether a surface water source for drinking water was contaminated (e.g., in Britain) was accomplished by placing a healthy fish into a stream. If it died, the source was contaminated and therefore must be purified. Arguments and comparisons over the degree of contamination were determined by a formula, calculated as 100 divided by the survival time in minutes. Our testing is more complex today, but sometimes not much more reliable.

While primarily designed as an information source, and presented in simple, straightforward, easy-to-understand English, *The Drinking Water Handbook* provides a level-headed look at a very serious situation, one based on years of extensive research on water quality. *The Drinking Water Handbook* is suitable for use by both the technical practitioner in the field and by students in the classroom. Here is all the information you need to make technical or personal decisions about drinking water.

CHAPTER 1

Introduction

When color photographs of the earth as it appears from space were first published, it was a revelation: they showed our planet to be astonishingly beautiful. We were taken by surprise. What makes the earth so beautiful is its abundant water. The great expanses of vivid blue ocean with swirling, sunlit clouds above them should not have caused surprise, but the reality exceeded everybody's expectations. The pictures must have brought home to all who saw them the importance of water to our planet. (E. C. Pielou, Fresh Water, Preface, 1998)

WATER is a contradiction, a riddle. Consider the Chinese proverb that states "water can both float and sink a boat."

Water's presence everywhere feeds these contradictions. Lewis (1996) points out that "water is the key ingredient of mother's milk and snake venom, honey and tears" (p. 90).

Leonardo da Vinci gave us insight into more of water's apparent contradictions:

Water is sometimes sharp and sometimes strong, sometimes acid and sometimes bitter;

Water is sometimes sweet and sometimes thick or thin;

Water sometimes brings hurt or pestilence, sometimes health-giving, sometimes poisonous.

Water suffers changes into as many natures as are the different places through which it passes.

Water, as with the mirror that changes with the color of its object, so it alters with the nature of the place, becoming: noisome, laxative, astringent, sulfurous, salt, incarnadined, mournful, raging, angry, red, yellow, green, black, blue, greasy, fat or slim.

Water sometimes starts a conflagration, sometimes it extinguishes one.

Water is warm and is cold.

Water carries away or sets down.

Water hollows out or builds up.

Water tears down or establishes.

Water empties or fills.

Water raises itself or burrows down.

Water spreads or is still.

Water is the cause at times of life or death, or increase of privation, nourishes at times and at others does the contrary.

Water, at times has a tang, at times it is without savor.

Water sometimes submerges the valleys with great flood.

In time and with water, everything changes.

We can sum up water's contradictions by simply stating that, though the globe is awash in it, water is no single thing, but an elemental force that shapes our existence. Da Vinci's last contradiction, "In time and with water, everything changes" concerns us most in this text.

We stated in the preface that next to the air we breathe, the water we drink is most important to us—to all of us. Water is no less important than air, simply less urgent. For all of us, though we treat it casually, unthinkingly, water is not a novelty, but a necessity—we simply cannot live without water.

Some might view our statements about the vital importance of water as nothing more than hyperbole. But are they? Is our concern over safe drinking water really an exaggeration? We think not, because we absolutely know and understand this simple point: We were born of water, and to live we must be sustained by it.

Development of safe drinking water supplies is a major concern today. This may seem strange to the average person; however, drinking water practitioners (those responsible for finding a source, certifying its safety, and providing it to the consumer) know better. The drinking water practitioner knows, for example, that two key concerns drive the development of safe drinking water supplies: Quantity and Quality (Q and Q). Quantity may indeed be a major issue (a limiting factor) for a particular location—often the case because water suitable for consumption is not evenly distributed throughout the world. Those locations that have an ample supply of surface water or groundwater may not have a "quantity" problem—as long as the quantity is large enough to fulfill the needs of all the consumers. But, again, not every geographical location is fortunate enough to have an adequate water supply—that is, the quantity of water available to satisfy residents' needs. This is one of the primary reasons, of course, that major portions of the globe are either uninhabited, or sparsely populated at best.

The other key concern (and the main focus of this text) is water quality. Obviously, having a sufficient quantity of freshwater available does little good if the water is unsafe for consumption or for other uses.

We began this chapter by pointing out the contradictions with water. At this point, we must point out another contradiction—one that human beings bear considerable responsibility for. Consider that most of the early settlements of the world began along waterways. Waterways were important primarily because of the ease of transportation they afforded and because of their industrial value (waterpower, etc.), as well as because of the food supply they provided to the settlers (fish and other wildlife). And of course, such waterways provided a natural, relatively clean, relatively safe source of drinking water.

However, these early waterways soon became polluted. Pollution is a natural by-product of civilization. We eat, we work, we do whatever is necessary to sustain our existence, and in doing so we pollute.

Are we the only freshwater polluter? No, not exactly. Natural occurrences also pollute our water sources, especially our surface waters. For example, a stream that flows in a heavily wooded area—in and through a deciduous forest, for example—suffers from the effects of natural pollution each year during leaf-fall.

When leaves fall from their lofty perches and make their sinuous descent into the blue-green phantoms we call surface streams below, they are carried with the flow, drifting until they sink, saturated, or lodge in an obstruction in the stream somewhere.

Leaves are organic and eventually degrade. During this process, the microbes degrading the leaves take up and use dissolved oxygen in the water. In some cases, the amount of oxygen used during degradation is of such quantity (especially in slow moving or stagnant water areas) that the natural biota in the stream suffer from a lack of oxygen and, therefore, either move on to healthier parts of the stream or succumb because of that lack. Note that other natural water polluters also affect water quality, including forest fires, earthquakes, and floods.

We pointed out many of the effects of natural pollutants in the opening statement about the contradictions of water. But remember that Nature understands the contradictions of water . . . and because she understands, she is also well-suited and equipped to deal with such problems. When a stream is polluted (for whatever reason), Nature immediately goes to work and sets in motion natural processes—self-purification—designed to restore the stream to its normal, healthy state.

Only when such streams become overloaded with pollution, or with nonbiodegradable man-made ingredients (contaminants), does Nature have difficulty in restoring the stream to its normal quality.

In the Preface, we stated that this text does not deal with the politics of drinking water pollution. Nor is this an environmental text. Instead, this text is intended for use as a handheld quick reference and technical support for the general public, sanitary engineers, public health administrators, public works engineers, water treatment operators, and college students in environmental health or public health engineering.

The purpose of this handbook is to evaluate and emphasize drinking water quality control, from the source to the treatment plant, from the distribution system to the consumer. The goal is to look at all the details, including trihalomethanes, *Cryptosporidium*, viruses, carcinogens, polychlorinated biphenols, etc., in addition to the traditional parameters—physical, chemical, and bacteriological. We discuss the overall view of the drinking water practitioner on providing the best drinking water quality to the consumer—the problems involved with providing a palatable and safe product for human consumption. Specifically, we deal with the nature of the problem and the solution—unsafe drinking water, and how to make it safe. And we also deal with the causes of the problem—pollution, and the techniques that can be employed to mitigate the problem through technology. We focus on science and technology because we have a clear understanding that technology and politics are seldom a rational mix.

1.1 REFERENCES

Lewis, S. A., *The Sierra Club Guide to Safe Drinking Water*. San Francisco: Sierra Club Books, 1996.

Pielou, E. C., *Fresh Water*. Chicago: University of Chicago Press, 1998.

CHAPTER 2

All About Water:
Basic Concepts

. . . During another visit to the New England Medical Center, three months after Robbie's first complaints of bone pain, doctors noted that his spleen was enlarged and that he had a decreased white-blood-cell count with a high percentage of immature cells—blasts—in the peripheral blood. A bone marrow aspiration was performed. The bone marrow confirmed what the doctors had begun to suspect: Robbie had acute lymphatic leukemia. (Jonathan Harr, A Civil Action, p. 30, 1995)

2.1 INTRODUCTION

TAKE a moment and perform an action most people never think about doing. Hold a glass of water (like the one in Figure 2.1) and consider the substance within the glass. We are aware that the water inside a drinking glass is not one of those items people usually spend much thought on, unless they are tasked with providing the drinking water.

Earlier we stated that water is special, strange, and different. We find water fascinating—a subject worthy of endless interest, because of its unique behavior, limitless utility, and ultimate and intimate connection with our existence. We agree with Tom Robbins, whose description of water follows.

Stylishly composed in any situation—solid, gas or liquid—speaking in penetrating dialects understood by all things—animal, vegetable or mineral—water travels intrepidly through four dimensions, *sustaining* (Kick a lettuce in the field and it will yell "Water!") *destroying* (The Dutch boy's finger remembered the view from Ararat) and *creating* (It has even been said that human beings were invented by water as a device for transporting itself from one place to another, but that's another story). Always in motion, ever-flowing (whether at stream rate or glacier speed), rhythmic, dynamic, ubiquitous, changing and working its changes, a mathematics turned wrong side out, a philosophy in reverse, the ongoing odyssey of water is irresistible. (*Even Cowgirls Get the Blues*, pp. 1–2, 1976).

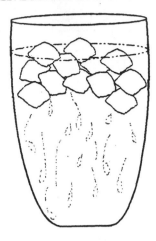

Figure 2.1

Let's review a few basic facts about the water in the glass you're holding: Water is liquid between 0°C and 100°C (32°F and 212°F), solid at or below 0°C (32°F), and gaseous at or above 100°C (212°F).

One gallon of water weighs 8.33 pounds (3.778 kilograms). One gallon of water equals 3.785 liters. One cubic foot of water equals 7.50 gallons (28.35 liters). One ton of water equals 240 gallons. One acre foot water equals 43,560 cubic feet (325,900 gallons). Earth's rate of rainfall equals 340 cubic miles per day (16 million tons per second).

As Robbins said, water is always in motion. The one most essential characteristic of water is that it is dynamic: Water constantly evaporates from sea, lakes, and soil and transpires from foliage; is transported through the atmosphere; falls to Earth; runs across the land; and filters down to flow along rock strata into aquifers. Eventually water finds its way to the sea again—indeed, water never stops moving.

A thought that might not have occurred to most people as we look at our glass of water is, "Who has tasted this same water before us?" Before us? Absolutely. Remember, water is a finite entity. What we have now is what we have had in the past. The same water consumed by Cleopatra, Aristotle, da Vinci, Napoleon, Joan of Arc (and several billion other folks who preceded us), we are drinking now—because water is dynamic (never at rest), and because water constantly cycles and recycles, as we discuss in the next section.

2.2 THE WATER CYCLE

The natural *water cycle* or *hydrological cycle* is the means by which water in all three forms—solid, liquid, and vapor—circulates through the biosphere. Water, lost from the Earth's surface to the atmosphere either by evaporation from the surface of lakes, rivers, and oceans or through the transpiration of plants, forms clouds that condense to deposit moisture on the land and sea. A drop of water may travel thousands of miles between the time it evaporates and the time it falls to Earth again as rain, sleet, or snow. The water that collects on land flows to the ocean in streams and rivers or seeps into the earth, joining groundwater. Even groundwater eventually flows toward the ocean for recycling (see Figure 2.2).

Note: Only about 2% of the water absorbed into plant roots is used in photosynthesis. Nearly all of it travels through the plant to the leaves, where transpiration to the atmosphere begins the cycle again.

Key Concept: The hydrologic cycle describes water's circulation through

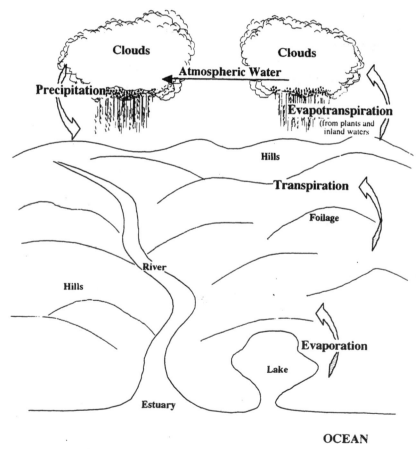

Figure 2.2 Water cycle: Natural. *Source:* Adapted from Hammer and Hammer, p. 2, 1996.

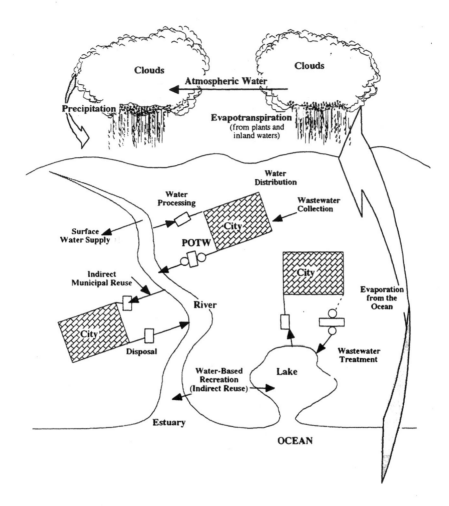

Figure 2.3 Integrated water cycle: Natural and human generated. *Source:* Adapted from Hammer and Hammer, p. 2, 1996.

the environment. Evaporation, transpiration, runoff, and precipitation describe specific water movements.

When humans intervene in the natural water cycle, they generate artificial water cycles or *urban* water cycles (local sub-systems of the water cycle—an integrated water cycle; see Figure 2.3). While many communities withdraw groundwater for public supply, the majority rely on surface sources. After treatment, water is distributed to households and industries. Water that is wasted (wastewater) is collected in a sewer system and transported to a treatment plant for processing prior to disposal. Current processing technologies provide only partial recovery of the original water quality. The upstream community (the first water user, shown in Figure 2.3) is able to achieve additional quality improvement by dilution into a surface water body and natural purification. However (as also shown in Figure 2.3), the next community downstream is likely to withdraw the water for a drinking water supply before complete restoration. This practice is intensified and further complicated as existing communities continue to grow, and new communities spring up along the same watercourse. Obviously, increases in the number of users bring additional need for increased quantities of water. This withdrawal and return process by successive communities in a river basin results in *indirect water reuse*.

Note: Metcalf and Eddy, Inc. (1991) define indirect water reuse as the potable reuse by incorporation of reclaimed wastewater into a raw water supply; the wastewater effluent is discharged to the water source, where it mixes with the source water and is assimilated with it, with the intent of reusing the water instead of as a means of disposal. This type of potable reuse is becoming more common as water resources become less plentiful (pp. 1139–1140).

As Hammer and Hammer (1996) point out, the indirect water reuse process (demonstrated in Figure 2.3) is a clear example of an artificial water cycle with the natural hydrologic scheme, involving (1) surface-water withdrawal, processing, and distribution; (2) wastewater collection, treatment, and disposal back to surface water by dilution; (3) natural purification in a river; and (4) repetition of this scheme by communities downstream (p. 1).

2.3 WATER SUPPLY: THE Q AND Q FACTORS

While drinking water practitioners must have a clear and complete understanding of the natural and man-made water cycles, they must also factor in two major considerations (Quantity and Quality–the Q and Q factors): (1) providing a "quality" potable water supply, one that is clean, wholesome, and safe to drink; and (2) finding a water supply available in adequate "quantities" to meet the anticipated demand.

Note: Two central facts important to our discussion of freshwater supplies are: (1) water is very much a local or regional resource, and (2) problems of water shortage or pollution are equally local problems. Human activities affect the quantity of water available at a locale at any time by changing either the total volume that exists there, or aspects of quality that restrict or devalue it for a particular use. Thus, the total human impact on water supplies is the sum of the separate human impacts on the various drainage basins and groundwater aquifers. In the global system, the central, critical fact about water is the natural variation in its availability (Meyer, 1996).

To meet the Q and Q requirements of a potential water supply, the drinking water practitioner (whether the design engineer, community planner, plant manager, plant administrator, plant engineer, or other responsible person in charge) must determine the answers to a number of questions, including:

(1) Does a potable water supply with the capacity to be distributed in sufficient quantity and pressure at all times exist nearby?

(2) Will constructing a centralized treatment and distribution system for the whole community be best, or would using individual well supplies be better?

(3) If a centralized water treatment facility is required, will the storage capacity at the source as well as at intermediate points of the distribution system maintain the water pressure and flow (quantity) within the conventional limits, particularly during loss-of-pressure "events"—major water main breaks, rehabilitation of the existing system, or major fires, for example?

(4) Is a planned or preventive maintenance program in place (or anticipated) for the distribution system that can be properly planned, implemented, and controlled at the optimum level possible?

(5) Is the type of water treatment process selected in compliance with federal and state drinking water standards?

Note: Water from a river or a lake usually requires more extensive treatment than groundwater does, to remove bacteria and suspended particles.

Note: The primary concern for the drinking water practitioner involved with securing an appropriate water supply, treatment process, and distribution system must be the protection of public health. Contaminants must be eliminated or reduced to a safe level to minimize menacing waterborne diseases (to prevent another Milwaukee *Cryptosporidium* event—see Chapter 6) and to avoid forming long-term or chronic injurious health effects.

(6) Once the source and treatment processes are selected, has the optimum hydraulic design of the storage, pumping, and distribution network been determined to ensure that sufficient quantities of water can be delivered to consumers at adequate pressures?

(7) Have community leaders and the consumer (the general public) received continuing and realistic information about the functioning of the proposed drinking water service?

Note: Drinking water practitioners are wise to follow the guidance given in (7), simply because the public buy-in to any proposed drinking water project that involves new construction and/or retrofitting, expansion, or upgrade of an existing facility is essential—to ensure that necessary financing is forthcoming. In addition to the finances needed for any type of waterworks construction project, public and financial support is also required to ensure the safe operation, maintenance, and control of the entire water supply system. The acronym POTW (Publicly Owned Treatment Works) begins with the word "public," and the public foots the bills.

(8) Does planning include steps to ensure elimination of waste, leakages, and unauthorized consumption?

Note: Industry-wide operational experience has shown that the cost per cubic foot, cubic meter, liter, or gallon of water delivered to the customer has steadily increased because of manpower, automation, laboratory, and treatment costs. To counter these increasing costs, treatment works must meter consumers, measure the water supply flow, and should evaluate the entire system annually.

(9) Does the waterworks or proposed waterworks physical plant include adequate laboratory facilities to ensure proper monitoring of water quality?

Note: Some waterworks facilities routinely perform laboratory work. However, water pollution control technologists must ensure that the waterworks laboratory or other laboratory used is approved by the appropriate health authority. Keep in mind that the laboratory selected to test and analyze the waterworks samples must be able to analyze chemical, microbiologic, and radionuclide parameters.

(10) Are procedures in place to evaluate specific problems such as the lead content in the distribution systems and at the consumer's faucet, or suspected contamination due to cross-connection potentials?

(11) Is a cross-connection control program in place to make sure that the distribution system (in particular) is protected from plumbing errors and illegal connections that may lead to injection of nonpotable water into public or private supplies of drinking water?

(12) Are waterworks operators and laboratory personnel properly trained and licensed?

(13) Are waterworks managers properly trained and licensed?

(14) Are proper operating records and budgetary records kept?

2.4 DRINKING WATER QUALITY AND QUANTITY: KEY DEFINITIONS

As with any other technical presentation, understanding the information presented is difficult unless a common vocabulary is established. Voltaire said it best. "If you wish to converse with me, please define your terms." We define many key terms used in the text in glossary form in this section. We define other terms as they appear in the text. Before we list and define these key terms, however, we must first define "the" key term that this text is all about: *drinking water*.

Drinking water or *potable* water can be defined as the water delivered to the consumer that can be safely used for drinking, cooking, washing, and other household applications. In the past, drinking water suitable for safe use was simply certified as safe by a professional engineer specialized in the field—but times have changed and this practice is no longer accepted. Why? Because public health aspects have reached such a high level of importance and complexity that local licensed health officials usually must be designated as those with the authority and jurisdiction in the community to review, inspect, sample, monitor, and evaluate the water supplied to a community on a continuing basis. This professional scrutiny is driven by updated drinking water standards. When you factor in the importance of providing a safe, palatable product to the public, requiring public health control to help ensure and guarantee a continuous supply of safe drinking water makes real sense.

On many occasions, we (as drinking water practitioners) have attempted to explain to interested parties the complexities of providing safe drinking water to their household taps. Of course, we know the process is complex, but after our initial surprise when listeners were astonished at the complex procedures and processes involved, we've come to expect that response. We see signs of a commonly held view that the provision of drinking water to the consumer tap involves nothing more than going down to the local river or stream, installing a suction pipe into the watercourse, and pumping the water out and into a distribution network that eventually (magically) delivers clean, safe drinking water to the household tap.

Under such circumstances (more common than you might imagine), we explain that, indeed, water is often taken from a local river or stream, and that it does eventually find its way to the household tap. However, we also explain that the water drawn from any local surface water supply goes through certain processes to ensure wholesomeness and safety. Aside from the physical treatment processes to screen, filter, and disinfect the water, these processes include inspecting, sampling, monitoring, and evaluating the water supply continuously.

Drinking water practitioners learn the ins and outs of drinking water, mainly from experience. The main lesson learned: Supplying drinking water to the household tap is a complex and demanding process. For example, water analysis is required (obviously), but it alone is not sufficient to maintain quality. It must be combined with the periodic review and acceptance of the facilities involved.

Acceptance or approval consists of the evaluation and maintenance of proper protection of the water source, qualifications of waterworks' personnel, water purveyor's (supplier's) adequate monitoring procedures, and also evaluation of the quality and performance of laboratory work (De Zuane, 1997).

Thus, when we attempt to define "drinking" water, we must define it in all-encompassing terms. Drinking or potable water is a product that meets the physical, chemical, bacteriological, and radionuclide parameters from an approved source, delivered to a treatment works for processing and disinfecting. Such a treatment works must be properly designed, constructed, and operated. Drinking water must be delivered to the consumer in sufficient quantity and pressure. Drinking water quality must meet stringent standards. It must be palatable, be within reasonable temperature limits, and have the complete confidence of the consumer.

The bottom line is that drinking water is that substance available to the consumer at the household tap that, when drawn from the tap, can perform one essential function: It can satisfy thirst without threatening life and health.

2.4.1 DEFINITIONS

- *Absorption:* any process by which one substance penetrates the interior of another substance.
- *Acid rain:* precipitation with higher than normal acidity, caused primarily by sulfur and nitrogen dioxide air pollution.
- *Activated carbon:* a very porous material that after being subjected to intense heat to drive off impurities can be used to adsorb pollutants from water.
- *Adsorption:* the process by which one substance is attracted to and adheres to the surface of another substance, without actually penetrating its internal structure.
- *Aeration:* a physical treatment method that promotes biological degradation of organic matter. The process may be passive (when waste is exposed to air), or active (when a mixing or bubbling device introduces the air).
- *Aerobic bacteria:* a type of bacteria that requires free oxygen to carry out metabolic function.
- *Biochemical Oxygen Demand (BOD):* the amount of oxygen required by bacteria to stabilize decomposable organic matter under aerobic conditions.
- *Biological treatment:* a process that uses living organisms to bring about chemical changes.
- *Breakpoint chlorination:* the addition of chlorine to water until the chlorine demand has been satisfied and free chlorine residual is available for disinfection.
- *Chemical treatment:* a process that results in the formation of a new

substance or substances. The most common chemical water treatment processes include coagulation, disinfection, water softening, and filtration.

- *Chlorination:* the process of adding chlorine to water to kill disease-causing organisms or to act as an oxidizing agent.
- *Chlorine demand:* a measure of the amount of chlorine that will combine with impurities and is therefore unavailable to act as a disinfectant.
- *Clean Water Act (CWA):* federal law dating to 1972 (with several amendments) with the objective to restore and maintain the chemical, physical, and biological integrity of the nation's waters. Its long-range goal is to eliminate the discharge of pollutants into navigable waters, and to make national waters fishable and swimmable.
- *Coagulants:* chemicals that cause small particles to stick together to form larger particles.
- *Coagulation:* a chemical water treatment method that causes very small suspended particles to attract one another and form larger particles. This is accomplished by the addition of a coagulant that neutralizes the electrostatic charges that cause particles to repel each other.
- *Coliform bacteria:* a group of bacteria predominantly inhabiting the intestines of humans or animals, but also occasionally found elsewhere. Presence of the bacteria in water is used as an indication of fecal contamination (contamination by animal or human wastes).
- *Color:* a physical characteristic of water. Color is most commonly tan or brown from oxidized iron, but contaminants may cause other colors, such as green or blue. Color differs from turbidity, which is water's cloudiness.
- *Communicable diseases:* usually caused by microbes—microscopic organisms including bacteria, protozoa, and viruses. Most microbes are essential components of our environment and do not cause disease. Those that do are called pathogenic organisms, or simply *pathogens.*
- *Community water system:* a public water system that serves at least 15 service connections used by year-round residents, or regularly serves at least 25 year-round residents.
- *Composite sample:* a series of individual or grab samples taken at different times from the same sampling point and mixed together.
- *Contaminant:* a toxic material found as an unwanted residue in or on a substance.
- *Cross connection:* any connection between safe drinking water and a nonpotable water or fluid.
- $C \times T$ *value:* the product of the residual disinfectant concentration C, in milligrams per liter, and the corresponding disinfectant contact time

T, in minutes. Minimum $C \times T$ values are specified by the Surface Water Treatment Rule, as a means of ensuring adequate kill or inactivation of pathogenic microorganisms in water.

- *Disinfectants-Disinfection By-Products (D-DBPs):* a term used in connection with state and federal regulations designed to protect public health by limiting the concentration of either disinfectants or the by-products formed by the reaction of disinfectants with other substances in the water (such as trihalomethanes—THMs).
- *Disinfection:* a chemical treatment method. The addition of a substance (e.g., chlorine, ozone, or hydrogen peroxide) that destroys or inactivates harmful microorganisms, or inhibits their activity.
- *Dissociate:* the process of ion separation that occurs when an ionic solid is dissolved in water.
- *Dissolved Oxygen (DO):* the oxygen dissolved in water usually expressed in milligrams per liter, parts per million, or percent of saturation.
- *Dissolved solids:* any material that can dissolve in water and be recovered by evaporating the water after filtering the suspended material.
- *Drinking water standards:* water quality standards measured in terms of suspended solids, unpleasant taste, and microbes harmful to human health. Drinking water standards are included in state water quality rules.
- *Drinking water supply:* any raw or finished water source that is or may be used as a public water system or as drinking water by one or more individuals.
- *Effluent limitations:* standards developed by the EPA to define the levels of pollutants that could be discharged into surface waters.
- *Electrodialysis:* the process of separating substances in a solution by dialysis, using an electric field as the driving force.
- *Electronegativity:* the tendency for atoms that do not have a complete octet of electrons in their outer shell to become negatively charged.
- *Enhanced Surface Water Treatment Rule (ESWTR):* a revision of the original Surface Water Treatment Rule that includes new technology and requirements to deal with newly identified problems.
- *Facultative bacteria:* a type of anaerobic bacteria that can metabolize its food either aerobically or anaerobically.
- *Federal Water Pollution Control Act (1972):* the Act outlines the objective "to restore and maintain the chemical, physical, and biological integrity of the nation's waters." This 1972 act and subsequent Clean Water Act amendments are the most far-reaching water pollution control legislation ever enacted. They provided for comprehensive programs for water pollution control, uniform laws, and interstate cooperation. They provided grants for research,

investigations, training, and information on national programs on surveillance, the effects of pollutants, pollution control, and the identification and measurement of pollutants. Additionally, they allotted grants and loans for the construction of treatment works. The Act established national discharge standards with enforcement provisions.

The Federal Water Pollution Control Act established several milestone achievement dates. It required secondary treatment of domestic waste by publicly owned treatment works (POTWs) and application of "best practicable" water pollution control technology by industry by 1977. Virtually all industrial sources have achieved compliance. (Because of economic difficulties and cumbersome federal requirements, certain POTWs obtained an extension to July 1, 1988 for compliance.) The Act also called for new levels of technology to be imposed during the 1980s and 1990s, particularly for controlling toxic pollutants.

The Act mandated a strong pretreatment program to control toxic pollutants discharged by industry into POTWs. The 1987 amendments required that stormwater from industrial activity must be regulated.

- *Filtration:* a physical treatment method for removing solid (particulate) matter from water by passing the water through porous media such as sand or a man-made filter.
- *Flocculation:* the water treatment process following coagulation; it uses gentle stirring to bring suspended particles together so that they will form larger, more settleable clumps called floc.
- *Grab sample:* a single water sample collected at one time from a single point.
- *Groundwater:* the freshwater found under the Earth's surface, usually in aquifers. Groundwater is a major source of drinking water, and the quantity of groundwater is a growing concern in areas where leaching agricultural or industrial pollutants or substances from leaking underground storage tanks are contaminating it.
- *Hardness:* a characteristic of water caused primarily by the salts of calcium and magnesium. It causes deposition of scale in boilers, damage in some industrial processes, and sometimes objectionable taste. It may also decrease soap's effectiveness.
- *Hydrogen bonding:* the term used to describe the weak but effective attraction that occurs between polar covalent molecules.
- *Hydrologic cycle:* literally the water-earth cycle. The movement of water in all three physical forms through the various environmental mediums (air, water, biota, and soil).
- *Hygroscopic:* a substance that readily absorbs moisture.
- *Influent:* water flowing into a reservoir, basin, or treatment plant.

- *Inorganic chemical:* a chemical substance of mineral origin not having carbon in its molecular structure.
- *Ionic bond:* the attractive forces between oppositely charged ions—for example, the forces between the sodium and chloride ions in a sodium chloride crystal.
- *Maximum Contaminant Level (MCL):* the maximum allowable concentration of a contaminant in drinking water, as established by state and/or federal regulations. Primary MCLs are health related and mandatory. Secondary MCLs are related to the aesthetics of the water and are highly recommended, but not required.
- *Membrane filter method:* a laboratory method used for coliform testing. The procedure uses an ultrathin filter with a uniform pore size smaller than bacteria (less than a micron). After water is forced through the filter, the filter is incubated in a special media that promotes the growth of coliform bacteria. Bacterial colonies with a green-gold sheen indicate the presence of coliform bacteria.
- *Modes of transmission of disease:* the ways in which diseases spread from one person to another.
- *Multiple-tube fermentation method:* a laboratory method used for coliform testing, which uses a nutrient broth placed in a culture tube. Gas production indicates the presence of coliform bacteria.
- *National Primary Drinking Water Regulations (NPDWRs):* regulations developed under the Safe Drinking Water Act, which establish maximum contaminant levels, monitoring requirements, and reporting procedures for contaminants in drinking water that endanger human health.
- *National Pollutant Discharge Elimination System (NPDES):* a requirement of the CWA that discharges meet certain requirements prior to discharging waste to any water body. It sets the highest permissible effluent limits, by permit, prior to making any discharge.
- *Near Coastal Water Initiative:* this initiative was developed in 1985 to provide for management of specific problems in waters near coastlines that are not dealt with in other programs.
- *Nonbiodegradable:* substances that do not break down easily in the environment.
- *Nonpolar covalently bonded:* a molecule composed of atoms that share their electrons equally, resulting in a molecule that does not have polarity.
- *Organic chemical:* a chemical substance of animal or vegetable origin having carbon in its molecular structure.
- *Oxidation:* when a substance either gains oxygen or loses hydrogen or electrons in a chemical reaction. One of the chemical treatment methods.

- *Oxidizer:* a substance that oxidizes another substance.
- *Parts Per Million:* the number of weight or volume units of a constituent present with each one million units of the solution or mixture. Formerly used to express the results of most water and wastewater analyses, *PPM* is being replaced by milligrams per liter *M/L*. For drinking water analyses, concentration in parts per million and milligrams per liter are equivalent. A single PPM can be compared to a shot glass full of water inside a swimming pool.
- *Pathogens:* types of microorganisms that can cause disease.
- *Physical treatment:* any process that does not produce a new substance (e.g., screening, adsorption, aeration, sedimentation, and filtration).
- *Polar covalent bond:* the shared pair of electrons between two atoms are not equally held. Thus, one of the atoms becomes slightly positively charged and the other atom becomes slightly negatively charged.
- *Polar covalent molecule:* one or more polar covalent bonds result in a molecule that is polar covalent. Polar covalent molecules exhibit partial positive and negative poles, causing them to behave like tiny magnets. Water is the most common polar covalent substance.
- *Pollutant:* any substance introduced into the environment that adversely affects the usefulness of the resource.
- *Pollution:* the presence of matter or energy whose nature, location, or quantity produces undesired environmental effects. Under the Clean Water Act, for example, the term is defined as a man-made or man-induced alteration of the physical, biological, and radiological integrity of water.
- *Pretreatment:* any physical, chemical, or mechanical process used before the main water treatment processes. It can include screening, presedimentation, and chemical addition.
- *Primary Drinking Water Standards:* regulations on drinking water quality (under SDWA) considered essential for preservation of public health.
- *Primary treatment:* the first step of treatment at a municipal wastewater treatment plant. It typically involves screening and sedimentation to remove materials that float or settle.
- *Public water system:* as defined by the Safe Drinking Water Act, any system, publicly or privately owned, that serves at least 15 service connections 60 days out of the year or serves an average of 25 people at least 60 days out of the year.
- *Publicly Owned Treatment Works (POTW):* a waste treatment works owned by a state, local government unit, or Indian tribe, usually designed to treat domestic wastewaters.
- *Receiving waters:* a river, lake, ocean, stream, or other water source into which wastewater or treated effluent is discharged.

- *Recharge:* the process by which water is added to a zone of saturation, usually by percolation from the soil surface.
- *Reference Dose (RfD):* an estimate of the amount of a chemical that a person can be exposed to on a daily basis that is not anticipated to cause adverse systemic health effects over the person's lifetime.
- *Representative sample:* a sample containing all the constituents present in the water from which it was taken.
- *Reverse Osmosis (RO):* solutions of differing ion concentration are separated by a semipermeable membrane. Typically, water flows from the chamber with lesser ion concentration into the chamber with the greater ion concentration, resulting in hydrostatic or osmotic pressure. In RO, enough external pressure is applied to overcome this hydrostatic pressure, thus reversing the flow of water. This results in the water on the other side of the membrane becoming depleted in ions and demineralized.
- *Safe Drinking Water Act (SDWA):* a federal law passed in 1974 with the goal of establishing federal standards for drinking water quality, protecting underground sources of water, and setting up a system of state and federal cooperation to assure compliance with the law.
- *Screening:* a pretreatment method that uses coarse screens to remove large debris from the water to prevent clogging of pipes or channels to the treatment plant.
- *Secondary Drinking Water Standards:* regulations developed under the Safe Drinking Water Act that established maximum levels of substances affecting the aesthetic characteristics (taste, color, or odor) of drinking water.
- *Secondary treatment:* the second step of treatment at a municipal wastewater treatment plant. This step uses growing numbers of microorganisms to digest organic matter and reduce the amount of organic waste. Water leaving this process is chlorinated to destroy any disease-causing microorganisms before its release.
- *Sedimentation:* a physical treatment method that involves reducing the velocity of water in basins so that the suspended material can settle out by gravity.
- *Solvated:* when either a positive or negative ion becomes completely surrounded by polar solvent molecules.
- *Surface tension:* the attractive forces exerted by the molecules below the surface upon those at the surface, resulting in them crowding together and forming a higher density.
- *Surface water:* all water naturally open to the atmosphere, and all springs, wells, or other collectors that are directly influenced by surface water.
- *Surface Water Treatment Rule (SWTR):* a federal regulation established by the USEPA under the Safe Drinking Water Act that

imposes specific monitoring and treatment requirements on all public drinking water systems that draw water from a surface water source.

- *Synthetic Organic Chemicals (SOCs):* generally applied to manufactured chemicals that are not as volatile as volatile organic chemicals. Included are herbicides, pesticides, and chemicals widely used in industries.
- *Total Suspended Solids (TSS):* solids present in wastewater.
- *Trihalomethanes (THMs):* a group of compounds formed when natural organic compounds from decaying vegetation and soil (such as humic and fulvic acids) react with chlorine.
- *Turbidity:* a measure of the cloudiness of water caused by the presence of suspended matter, which shelters harmful microorganisms and reduces the effectiveness of disinfecting compounds.
- *Vehicle of disease transmission:* any nonliving object or substance contaminated with pathogens.
- *Wastewater:* the spent or used water from individual homes, a community, a farm, or an industry that contains dissolved or suspended matter.
- *Waterborne disease:* water is a potential vehicle of disease transmission, and waterborne disease is possibly one of the most preventable types of communicable illness. The application of basic sanitary principles and technology have virtually eliminated serious outbreaks of waterborne diseases in developed countries. The most prevalent waterborne diseases include typhoid fever, dysentery, cholera, infectious hepatitis, and gastroenteritis.

   Note: Waterborne diseases are also called *intestinal diseases*, because they affect human intestinal tracts. If pathogens excreted in the feces of infected people are inadvertently ingested by others (in contaminated water, for example), the cycle of disease can continue, possibly in epidemic proportions. Symptoms of intestinal disease include diarrhea, vomiting, nausea, and fever. Intestinal diseases can incapacitate large numbers of people in an epidemic, and sometimes result in the deaths of many infected individuals. Water contaminated with untreated sewage is generally the most common cause of this type of disease (Nathanson, 1997). According to Gerba (1996), in practice, "hazard identification in the case of pathogens is complicated because several outcomes—from asymptomatic infections to death—are possible, and their outcomes depend upon the complex integration between the agent and the host. This interaction, in turn, depends on the characteristics of the host as well as the nature of the pathogen. Host factors, for example, include preexisting immunity, age, nutrition, ability to mount an immune response, and other nonspecific host factors. Agent factors include types and strains of the organism as well as its capacity to elicit an immune response" (pp. 357–358).

- *Water softening:* a chemical treatment method that uses either chemicals to precipitate or a zeolite to remove those metal ions (typically Ca^{2+}, Mg^{2+}, Fe^{3+}) responsible for hard water.
- *Watershed:* the land area that drains into a river, river system, or other body of water.
- *Wellhead protection:* the protection of the surface and subsurface areas surrounding a water well or wellfield supplying a public water system from contamination by human activity.

2.5 CLEAN, FRESH, AND PALATABLE: A HISTORICAL PERSPECTIVE

An early human, wandering alone from place to place, hunting and gathering to subsist, probably would have had little difficulty in obtaining drinking water, because such a person would—and could—only survive in an area where drinking water was available with little travail.

The search for clean, fresh, and palatable water has been a human priority from the very beginning. We take no risk in stating that when humans first walked the Earth, many of the steps they took were in the direction of water supply.

When early humans were alone or in small numbers, finding drinking water was a constant priority, to be sure, but for us to imagine today just how big a priority finding drinking water became as the number of humans proliferated is difficult.

Eventually communities formed, and with their formation came the increasing need to find clean, fresh, and palatable drinking water, and also to find a means of delivering it from the source to the point of use.

Archeological digs are replete with the remains of ancient water systems (man's early attempts to satisfy that never-ending priority). Those digs (spanning the history of the last 20 or more centuries) testify to this. For well over 2000 years, piped water supply systems have been in existence. Whether the pipes were fashioned from logs or clay, or carved from stone or other materials is not our point—the point is they were fashioned to serve a vital purpose, one universal to the community needs of all humans: to deliver clean, fresh, and palatable water to where it was needed.

These early systems were not arcane. We readily understand their intended purpose today. As we might expect, they could be rather crude, but they were reasonably effective, though they lacked in two general areas we take for granted today.

First, of course, they were not pressurized, but instead relied on gravity flow, since the means to pressurize the mains was not known at the time; and even if such pressurized systems were known, they certainly would not have been used to pressurize water delivered via hollowed-out logs and clay pipe.

The second general area early civilizations lacked that we do not today (in the industrialized world, that is) is sanitation. Remember, to know that the need for something exists (in this case, the ability to sanitize, to disinfect water supplies), the nature of the problem must be defined. Not until the middle of the 1800s (after countless millions of deaths from waterborne diseases over the centuries) did people realize that a direct connection between contaminated drinking water and disease exists. At that point, sanitation of water supply became an issue.

When the relationship between waterborne diseases and the consumption of drinking water was established, evolving scientific discoveries led the way toward the development of technology for processing and disinfection. Drinking water standards were developed by health authorities, scientists, and sanitary engineers.

With the current lofty state of effective technology that we in the U.S. and the rest of the developed world enjoy today, we could sit on our laurels, so to speak, and assume that because of the discoveries developed over time (and at the cost of countless people who died and still die from waterborne diseases), that all is well with us—that problems related to providing us with clean, fresh, palatable drinking water are problems of the past.

Are they really problems of the past? Have we solved all the problems related to ensuring that our drinking water supply provides us with clean, fresh, and palatable drinking water? Is the water delivered to our tap as clean, fresh, and palatable as we think it is . . . as we hope it is? Does anyone really know?

What we do know is that we have made progress. We have come a long way from the days of gravity flow water delivered via mains of logs and clay or stone . . . Many of us on this earth have come a long way from the days of cholera epidemics.

However, to obtain a definitive answer to those questions, perhaps we should ask those who boiled their water for weeks on end in Sydney, Australia in the fall of 1998. Or better yet, we should speak with those who drank the "city water" in Milwaukee in 1993, or in 1994, in Las Vegas, Nevada—those who suffered and survived the onslaught of *Cryptosporidium,* from contaminated water out of their tap.

Or if we could, we should ask these same questions of a little boy named Robbie, who died of acute lymphatic leukemia, the probable cause of which is far less understandable to us: toxic industrial chemicals, unknowingly delivered to him via his local water supply.

2.6 REFERENCES

De Zuane, J., *Handbook of Drinking Water Quality*, 2nd Edition. New York: John Wiley and Sons, 1997.

Gerba, C. P., Risk Assessment. In *Pollution Science*, eds., Pepper, Gerba, and Brusseau. San Diego: Academic Press, 1996.

Hammer, M. J. and Hammer, M. J., Jr., *Water and Wastewater Technology*, 3rd Edition. Englewood Cliffs, NJ: Prentice Hall, 1996.

Harr, J., *A Civil Action*. New York: Vintage Books, 1995.

Metcalf and Eddy, Inc., *Wastewater Engineering: Treatment, Disposal, Reuse*. 3rd Edition. New York: McGraw-Hill, Inc., 1991.

Meyer, W. B., *Human Impact on Earth*. New York: Cambridge University Press, 1996.

Nathanson, J. A., *Basic Environmental Technology: Water Supply, Waste Management, and Pollution Control*. Upper Saddle River, NJ: Prentice Hall, 1997.

Robbins, T., *Even Cowgirls Get the Blues*. Boston: Houghton Mifflin Company, 1976.

CHAPTER 3

Drinking Water Regulations

*Drinking water regulations have undergone major and dramatic changes in the past two decades, and trends indicate that they will continue to become more stringent and complicated. It is important that all water system operators understand the basic reasons for having regulations, how they are administered, and why compliance with them is so essential. (*Water Quality, *2nd ed., AWWA, p. xiii, 1995)*

3.1 INTRODUCTION

REGULATIONS—why do we need them? Most of us have little trouble answering this question, having no taste for anarchy. We regulate ourselves and others for a variety of reasons, but in our attempts to do so, we generally strive to attain similar results. For example, most governments regulate their population to provide direction, to manage, to monitor, and to literally govern whatever it is they are attempting to regulate (including us). We also regulate to confine, to control, to limit, and to restrict ourselves within certain parameters to maintain the peace—with the goal of providing equal and positive social conditions for us all. Regulations are not foreign to us . . . we are literally driven by them from birth through our final interment. You could say that we are literally regulated to death.

Some regulations are straightforward. The 65 mph speed limit on some interstates is simple—the regulation establishes measurable limits. Other regulations are not so simple. For example, certain regulations designed to ensure the safe and correct operation of nuclear reactors are complex and difficult to meet. Whether straightforward or complex, however, enforcement presents special problems. As to the safe drinking water regulations, we can only hope that those in place to ensure our safety and health are more effectively enforced than that 65 mph speed limit.

In this chapter, we discuss U.S. federal regulations designed to protect our health and well-being: the so-called Drinking Water Regulations. Control of the quality of our drinking water is accomplished by establishing certain regulations, which in turn require compliance within an established set of guidelines or parameters. The guidelines are the regulations themselves; the parameters are the water quality factors important to providing drinking water that is safe and palatable.

3.2 SETTING THE STAGE

Consider what might be an absurd question: Why do we need to regulate water quality?

And the next question, maybe a bit more logical: Aren't we already regulated enough?

The first question requires a compound answer, the explanation of which we provide in this chapter—we hope it will clear the water, so to speak. The second question? Based on our view, we use another question to answer that question. When it comes to ensuring a safe and palatable drinking water supply, are we (or can we be) regulated enough?

In this text, we concentrate on answering the first question because (even though it requires a compound answer), it goes to the heart of our discussion—the necessity of providing safe and palatable drinking water to the user.

Again, why do we need to regulate water quality? Let's start at the beginning.

In the beginning (the ancient beginning), humans really had no reason to give water quality much of a thought. Normally, nearly any water supply available was only nominally naturally polluted. Exceptions existed of course. For example, if a prehistoric human flattened out on the ground alongside a watercourse to drink the water, he or she didn't ingest too much of it—as little of it as possible—if it was salty. At this point our intrepid (but thirsty) ancestor probably moved on to find another water source, one a bit more palatable.

Determining water's fitness to drink was a matter of sight, smell, and a quick taste. If these criteria were met, the water was used. A water supply might look perfectly clear, smell OK, and not taste all that bad either, so our early kinfolk could have gulped it down until full and satisfied. However, later that day the water could have caused him to become sick, very sick—sickened by waterborne pathogens that were residents of that perfectly clear, not too bad tasting water ingested a few hours earlier. Of course, early humans would not have had the foggiest idea what caused the sickness, but they would have become sick indeed.

Let's take a look at more recent times, at another scenario that helps illustrate the point that we are making here.

An excursion to the local stream can be a relaxing and enjoyable undertaking. On the other hand, when you arrive at the local stream, spread your blanket on the streambank and then look out upon the stream's flowing mass only to discover a parade of waste and discarded rubble bobbing along the stream's course and cluttering the adjacent shoreline and downstream areas, any feeling of relaxation or enjoyment is quickly extinguished. Further, the sickening sensation the observer feels is not lessened, but made worse as he gains closer scrutiny of the putrid flow. He easily recognizes the rainbow-colored shimmer of an oil slick, interrupted here and there by dead fish and floating refuse, and the slimy fungal growth that prevails. At the same time, the observer's sense of smell is alerted to the noxious conditions. Along with the fouled water and the stench of rot-filled air, the observer notices the ultimate insult and tragedy: The signs warn: "Danger—No Swimming or Fishing." The observer soon realizes that the stream before him is not a stream at all; it is little more than an unsightly drainage ditch. The observer has discovered what ecologists have known and warned about for years. That is, contrary to popular belief rivers and streams do not have an infinite capacity for pollution. (Spellman, p. 65, 1996)

This relatively recent scenario (ca. 1996) makes an important point for us: The qualities of water that directly affect our senses are first to disturb us. This certainly was the case with ancient humans, before the discovery of what causes disease and waterborne disease in particular.

Even before the mid-1850s, when in London, Dr. John Snow made the connection between disease and water (i.e., the waterborne disease cholera), rumblings were heard, for example, about the terribly polluted state of the Thames River. Dr. Snow's discovery of the "connection" between cholera and the drinking water pumped from the Broad Street pump (ingested by those who became ill or died) lit the fire of reform, and revulsion set in motion steps to clean up the water supply. Since Snow's discovery, many subsequent actions taken to clean up a particular water supply resulted from incidents related to public disgust with the sorry state of the watercourse.

For example, in the 1960s, the burgeoning environmental movement found many ready examples of the deplorable state (and vulnerability) of America's waters. In Cleveland, the Cuyahoga River burst into flames, so polluted was it with chemicals and industrial wastes; historic Boston Harbor was a veritable cesspool; raw sewage spewed into San Francisco Bay. A 1969 oil spill off scenic Santa Barbara, California, proved an especially telegenic disaster, with oil-soaked seals and pelicans and miles of hideously fouled beaches. These and other incidents were disturbing to many Americans and brought calls for immediate reform.

Awareness of the state of our environment was at an all-time high. The power of the words of a brilliant writer began a grassroots crusade of environmental action in many areas, a writer whose penetrating scientific views and poetic prose captured the imagination of the nation. Rachel Carson became the flag bearer for our environment. By making the connections between isolated incidents and revealing the connections between industry, research, and government, she brought the clear light of day into the dark abyss of environmental degradation, revealing widespread horrible environmental conditions, and the future they could lead to.

The public lost trust in government's and industry's ability to self-govern on choices between money and the benefits of a clean environment for us all. Industry and government's close connections and financial self-interest were revealed as poor criteria for determining realistic levels of environmental protection. With Rachel Carson's *Silent Spring* came the sobering awareness that environmental conditions and the prevailing governmental attitude demanded radical change. Individual incidents disturbed many Americans to the point that they demanded immediate reforms.

3.3 CLEAN WATER REFORM IS BORN

To understand the history (and thus the impetus) behind the reform movement generated to clean up our water supplies (recent past to present), we trace a chronology of some of the significant events precipitated by environmental organizations and citizen groups that have occurred since the mid-1960s.[1]

(1) Americans came face-to-face with the grim condition of the nation's waterways in 1969, when the industrial-waste-laden Cuyahoga River caught on fire. That same year, waste from food processing plants killed almost 30 million fish in Lake Thonotosassa, Florida.

(2) In 1972, Congress enacted the Clean Water Act (after having overridden President Nixon's veto). The passage of the Clean Water Act has been called "literally a life-or-death proposition for the Nation" (Sierra Club, 1997, p. 4). The Act set the goals of achieving water quality levels that are "fishable and swimmable" by 1983; receiving zero discharges of pollutants by 1985; and prohibiting the discharge of toxic pollutants in toxic amounts.

(3) In 1974, the Safe Drinking Water Act (SDWA) passed, requiring the USEPA to establish national standards for contaminants in drinking water systems, underground wells, and sole-source aquifers, as well as several other requirements (see Section 3.5).

(4) In 1984, an alliance of the Natural Resources Defense Council, the Sierra Club, and others successfully sued Phillips ECG, a New York industrial polluter that had dumped waste into the Seneca River. According to the

[1]Chronology adapted from "Clean Water Timeline," *The Planet*, Volume 4, October 8, 1997, San Francisco: Sierra Club.

Sierra Club's water committee chair, Samuel Sage, the case "tested the muscles of citizens against polluters under the Clean Water Act" (Sierra Club, 1997, p. 6). During this same time frame, the Clean Water Act reauthorization bill drew the wrath of environmental groups, who dubbed it the "Dirty Water Act" after lawmakers added last minute pork and weakened wetlands protection and industrial pretreatment provisions. Because of grassroots action, most of these pork provisions were dropped. 1984 also saw the highest environmental penalty to date—$70,000—imposed against Alcoa Aluminum in Messina, N.Y. (for polluting the St. Lawrence River), as a result of a suit filed by the Sierra Club.

(5) In 1986, Tip O'Neill, Speaker of the House of Representatives, stated that he would not let a Clean Water Act reauthorization bill on the floor without the blessing of environmental groups. Later, after the bill was crafted and passed by Congress, President Reagan vetoed the bill. Also in 1986, amendments to the Safe Drinking Water Act directed the USEPA to publish a list of drinking water contaminants that require legislation.

(6) In 1987, the Clean Water Act was reintroduced. It became law after Congress overrode President Reagan's veto. A new provision established the National Estuary Program.

(7) From 1995 to 1996, the House passed H.R. 961 (again dubbed the Dirty Water Act), which in some cases eliminated standards for water quality, wetlands protection, sewage treatment, and agricultural and urban run-off. The Sierra Club collected over one million signatures supporting the Environmental Bill of Rights and released "Danger on Tap," a report that showed polluter contributions to friends in Congress who wanted to gut the Clean Water Act. Due in part to these efforts, the bill was stopped in the Senate.

(8) In 1997, the USEPA reported that more than one-third of the country's rivers and half its lakes were still unfit for swimming or fishing. The Sierra Club successfully sued the USEPA to enforce Clean Water Act regulations in Georgia. The state was required to identify polluted waters and establish their pollution-load capacity. Similar suits were filed in other states. For example, in Virginia, Smithfield Foods was assessed a penalty of more than $12 million—the highest ever—for violating the Clean Water Act by discharging phosphorous and other hog waste products into a tributary of the Chesapeake Bay.

This chronology of events presents only a handful of the significant actions taken by Congress (with the helpful prodding and guidance provided by the Sierra Club and the National Resources Defense Council as well as others) in enacting legislation and regulations to protect our nation's waters. No law has been more important in furthering this effort than the Clean Water Act, which we discuss in the following section.

3.4 THE CLEAN WATER ACT[2]

Concern with the disease-causing pathogens residing in many of our natural waterways was not the initial initiative that got Joe or Nancy Citizen's attention about the condition and health of the country's waterways. The aesthetic qualities of the watercourses first stirred their attention. Americans, in general, have a strong emotional response to the beauty of nature and acted to prevent the pollution and degradation of our nation's waterways simply because many of us expect rivers, waterfalls, and mountain lakes to be "natural" and naturally beautiful—in the state they were intended to be, pure and clean.

Much of this emotional attachment to the environment was generated from sentimentality borne by popular literature and art of early 19th-century American writers and painters. From Longfellow's *Song of Hiawatha* to Mark Twain's *Huckleberry Finn* to the vistas of the Hudson River School of Winslow Homer, American culture abounds with expressions of this singularly strong attachment. As the saying goes: "Once attached, detachment is never easy."

Federal water pollution legislation dates back to the turn of the century, to the Rivers and Harbors Act of 1899, though the Clean Water Act stems from the Federal Water Pollution Control Act originally enacted in 1948 to protect surface waters such as lakes, rivers, and coastal areas. The Act was significantly expanded and strengthened in 1972 in response to growing public concern for serious and widespread water pollution problems. This 1972 legislation provided the foundation for our dramatic progress in reducing water pollution over the past 25 years. Amendments to the 1972 Clean Water Act were made in 1977, 1981, and 1987.

The Clean Water Act focuses on improving water quality by maintaining and restoring the physical, chemical, and biological integrity of the nation's waters. It provides a comprehensive framework of standards, technical tools, and financial assistance to address the many stressors that can cause pollution and adversely affect water quality, including municipal and industrial wastewater discharges, polluted runoff from urban and rural areas, and habitat destruction.

The Clean Water Act requires national performance standards for major industries (such as iron and steel manufacturing and petroleum refining) that

---

[2]Much of the information contained in this section is from USEPA Clean Water Act obtained from www.epa.gov, dated 5/23/1996.

provide a minimum level of pollution control based on the best technologies available. These national standards result in the removal of over a billion pounds of toxic pollution from our waters every year.

The Clean Water Act also establishes a framework whereby states and Indian tribes survey their waters, determine an appropriate use (such as recreation or water supply), then set specific water quality criteria for various pollutants to protect those uses. These criteria, together with the national industry standards, are the basis for permits that limit the amount of pollution that can be discharged to a water body. Under the National Pollutant Discharge Elimination System, sewage treatment plants and industries that discharge wastewater are required to obtain permits and to meet the specified limits in those permits.

Note: The CWA requires the USEPA to set effluent limitations. All dischargers of wastewaters to surface waters are required to obtain NPDES permits, which require regular monitoring and reporting.

The Clean Water Act also provides federal funding to help states and communities meet their clean water infrastructure needs. Since 1972, federal funding has provided more than $66 billion in grants and loans, primarily for building or upgrading sewage treatment plants. Funding is also provided to address another major water quality problem—polluted runoff from urban and rural areas.

Protecting valuable aquatic habitat—wetlands, for example—is another important component of this law. American waterways have suffered loss and degradation of biological habitat, a widespread cause of the decline in the health of aquatic resources. When Europeans colonized this continent, North America held approximately 221 million acres of wetlands. Today most wetlands are lost. Roughly 22 states have lost 50% or more of their original acreage of wetlands. Ten states have lost about 70% of their wetlands.

Note that the Clean Water Act sections dealing with wetlands have become extremely controversial. Though wetlands are among our nation's most fragile ecosystems and provide a valuable role in maintaining regional ecology and preventing flooding, while serving as home to numerous species of insects, birds, and animals, wetlands also possess potential expandable monetary value in the eye of private landowners and developers. Herein lies the major problem. Many property owners feel they are being unfairly penalized by a Draconian regulation that restricts their right to develop their own property.

Alternative methods that do not involve destroying the wetlands do exist. These methods include wetlands mitigation and mitigation banking. Since 1972, when the Clean Water Act was passed, permits from the Army Corps of Engineers are needed to work in wetland areas. To obtain these permits, builders must agree to restore, enhance, or create an equal number of wetland acres (generally in the same watershed) as those damaged or destroyed in the construction project. Landowners are given the opportunity to balance the adverse effects by replacing environmental values that are lost. This concept is known as *wetlands mitigation.*

Mitigation banking allows developers or public bodies that seek to build on wetlands to make payments to a "bank" for use in the enhancement of other wetlands at a designated location. The development entity purchases "credit" in the bank and transfers full mitigation responsibility to an agency or environmental organization that runs the bank. Environmental professionals design, construct, and maintain a specific natural area using these funds.

The Clean Water Act's history is much like that of the environmental movement itself. Once widely supported, buoyed by its initial success, the Clean Water Act has encountered increasingly difficult problems—polluted stormwater runoff, for example, and nonpoint source pollution, as well as unforeseen legalistic challenges like the debate on wetlands and property rights.

Unfortunately, the Clean Water Act is only part of its way to the goal. At least one-third of the U.S. rivers, one-half of the U.S. estuaries, and more than one-half of the lakes are still not safe for such uses as swimming or fishing. Thirty-one states reported toxins in fish exceeding the action levels set by the Food and Drug Administration (FDA). Every pollutant in an EPA study on chemicals in fish showed up in at least one location. Water quality is seen as deteriorated and viewed as the cause of the decreasing number of shellfish in the waters.

3.5 THE SAFE DRINKING WATER ACT

When we get the opportunity to travel the world, one of the first things we learn to ask is whether or not the water is safe to drink. Unfortunately, in most of the places in the world, the answer is "no." Masters (1991) points out that as much as 80% of all sickness in the world is attributable to inadequate water or sanitation. William C. Clark, in a speech in Racine, Wisconsin (April 1988), probably summed it up best: "If you could tomorrow morning make water clean in the world, you would have done, in one fell sweep, the best thing you could have done for improving human health by improving environmental quality." According to Morrison (1983), an estimated three-fourths of the population in Asia, Africa, and Latin America lack a safe supply of water for drinking, washing, and sanitation. Money, technology, education, and attention to the problem are essential for improving these statistics, to solving the problem this West African proverb succinctly states: "Filthy water cannot be washed."

Left alone, Nature provides for us. Left alone, Nature feeds us. Left alone, Nature refreshes and sustains us with untainted air. Left alone, Nature provides and cleans the water we need to ingest to survive. As Norse (1985) put it, "In every glass of water we drink, some of the water has already passed through fishes, trees, bacteria, worms in the soil, and many other organisms, including people . . . Living systems cleanse water and make it fit, among other things, for human consumption" (p. 3). Left alone, Nature performs at a level of efficiency and perfection we can't imagine. The problem, of course, is that our human populations have grown too large to allow Nature to be left alone.

Our egos allow us to think that humans are the real reason Nature exists at all. In our eyes, our infinite need for water is why Nature works its hydrologic cycle—to provide the constant supply of drinking water we need to sustain life. But the hydrologic cycle itself is unstoppable, human activity or not. Bangs and Kallen (1985) sum it up best: "Of all our planet's activities—geological movements, the reproduction and decay of biota, and even the disruptive propensities of certain species (elephants and humans come to mind)—no force is greater than the hydrologic cycle" (pp. 2–3).

Nature, through the hydrologic cycle, provides us with an endless (we hope) re-supply of water. However, we find that developing and maintaining an adequate supply of safe drinking water requires the coordinated efforts of scientists, technologists, engineers, planners, water plant operators, and regulatory officials. In this section, we concentrate on the regulations that have been put into place in the U.S. to ensure that the water supplies developed are protected and maintained safe, fresh, and palatable.

Legislation to protect drinking water quality in the U.S. (the nation's first water quality standards) began with the Public Health Service Act of 1912. With time, the Act evolved, but not until the passage of the Safe Drinking Water Act (SDWA) of 1974 (amended 1986, 1996) was federal responsibility extended beyond interstate carriers to include all community water systems serving 15 or more outlets, or 25 or more customers. Prompted by public concern over findings of harmful chemicals in drinking water supplies, the law established the basic federal-state partnership for drinking water used today. It focuses on ensuring safe water from public water supplies and on protecting the nation's aquifers from contamination.

Before we examine the basic tenets of SDWA, we must define several of the terms used in the Act.

3.5.1 SDWA DEFINITIONS[3]

- *Best Management Practices (BMP):* schedules of activities, prohibitions of practices, maintenance procedures, and other management practices to prevent or reduce the pollution of "waters of the United States."

[3]Taken from 40 CFR Part 122.2; SDWA, 1401; CWA, 502.

- *Contaminant:* any physical, chemical, biological, or radiological substance or matter in water.
- *Discharge of a pollutant:* any addition of any pollutant to navigable waters from any point source.
- *Exemption:* a document for water systems having technical and financial difficulty meeting national primary drinking water regulations effective for one year granted by the USEPA "due to compelling factors."
- *Maximum Contaminant Level (MCL):* the maximum permissible level of a contaminant in water which is delivered to any user of a public water system.
- *Maximum Contaminant Level Goal (MCLG):* the level at which no known or anticipated adverse effects on the health of persons occur and which allows an adequate margin of safety.
- *National Pollutant Discharge Elimination System (NPDES):* the national program for issuing, modifying, revoking and reissuing, terminating, monitoring and enforcing permits, and imposing and enforcing pretreatment requirements, under sections 307, 402, 318, and 405 of the Clean Water Act.
- *Navigable waters:* waters of the United States, including territorial seas.
- *Person:* an individual, corporation, partnership, association, state, municipality, commission, or political subdivision of a State, or any interstate body.
- *Point source:* any discernible, confined, and discrete conveyance, including but not limited to any pipe, ditch, channel, tunnel, conduit, well, discrete fissure, container, rolling stock, concentrated animal feeding operation, or vessel, or other floating craft, from which pollutants are or may be discharged. This term does not include agricultural stormwater discharges and return flows from irrigated agriculture.
- *Pollutant:* dredged soil, solid waste, incinerator residue, filter backwash, sewage, garbage, sewage sludge, munitions, chemical wastes, biological materials, radioactive materials (except those regulated under the Atomic Energy Act of 1954), heat, wrecked or discarded equipment, rock, sand, cellar dirt, and industrial, municipal, and agricultural waste discharged into water. It does not mean: (a) sewage from vessels; or (b) water, gas, or other material which is injected into a well to facilitate production of oil or gas, or water derived in association with oil and gas production or for disposal purposes is approved by authority of the State in which the well is located, and if the State determines that the injection or disposal will not result in the degradation of ground or surface water sources.
- *Public water system:* a system for the provision to the public of piped

water for human consumption, if such system has at least fifteen service connections or regularly serves at least twenty-five individuals.

- *Publicly Owned Treatment Works (POTW):* any device or system used in the treatment of municipal sewage or industrial wastes of a liquid nature which is owned by a "state" or "municipality." This definition includes sewer, pipes, or other conveyances only if they convey wastewater to a POTW providing treatment.
- *Recharge Zone:* the area through which water enters a sole or principal source aquifer.
- *Significant hazard to public health:* any level of contaminant that causes or may cause the aquifer to exceed any maximum contaminant level set forth in any promulgated National Primary Drinking Water Standard at any point where the water may be used for drinking purposes or which may otherwise adversely affect the health of persons, or which may require a public water system to install additional treatment to prevent such adverse effect.
- *Sole or principal source aquifer:* an aquifer that supplies 50% or more of the drinking water for an area.
- *Streamflow source zone:* the upstream headwaters area that drains into an aquifer recharge zone.
- *Toxic pollutants:* those pollutants . . . which after discharge and upon exposure, ingestion, inhalation, or assimilation into any organism, will, on the basis of the information available to the Administrator, cause death, disease, behavioral abnormalities, cancer, genetic mutations, physiological malfunctions, or physical deformations, in such organisms or their offspring.
- *Variance:* a document for water systems having technical and financial difficulty meeting national primary drinking water regulations which postpones compliance when the issuing of which "will not result in an unreasonable risk to health."
- *Waters of the United States:* (a) all waters which are currently used, were used in the past, or may be susceptible to use in interstate or foreign commerce, including all waters which are subject to the ebb and flow of the tide; (b) all interstate waters, including interstate "wetlands"; (c) all other waters such as interstate lakes, rivers, streams, . . . mudflats, sandflats, "wetlands," sloughs, prairie potholes, wet meadows, playa lakes, or natural ponds, the use, degradation, or destruction of which would affect . . . interstate or foreign commerce.
- *Wetlands:* those areas that are inundated or saturated by surface or groundwater at a frequency and duration sufficient to support . . . a prevalence of vegetation typically adapted for life in saturated soil conditions. Wetlands generally include swamps, marshes, bogs, and similar areas.

3.5.2 SDWA: SPECIFIC PROVISIONS

To ensure the safety of public water supplies, the Safe Drinking Water Act requires the USEPA to set safety standards for drinking water. Standards are now in place for over 80 different contaminants. The USEPA sets a maximum level for each contaminant; however, in cases where making this distinction is not economically or technologically feasible, the USEPA specifies an appropriate treatment technology instead. Water suppliers must test their drinking water supplies and maintain records to ensure quality and safety. Most states carry the responsibility for ensuring their public water supplies are in compliance with the national safety standards.

Provisions also authorize the USEPA to conduct basic research on drinking water contamination, to provide technical assistance to states and municipalities, and to provide grants to states to help them manage their drinking water programs.

To protect groundwater supplies, the law provides a framework for managing underground injection compliance. As part of that responsibility, the USEPA may disallow new underground injection wells based on concerns over possible contamination of a current or potential drinking water aquifer.

Each state is expected to administer and enforce the SDWA regulations for all public water systems. Public water systems must provide water treatment, ensure proper drinking water quality through monitoring, and provide public notification of contamination problems. The 1986 amendments to the SDWA significantly expanded and strengthened its protection of drinking water. Under the 1986 provisions, the SDWA required six basic activities:

- Establishment and enforcement of Maximum Contaminant Levels (MCLs): as stated earlier, these are the maximum levels of certain contaminants that are allowed in drinking water from public systems. Under the 1986 amendments, the USEPA has set numerical standards or treatment techniques for an expanded number of contaminants.
- Monitoring: the USEPA requires monitoring of all regulated and certain unregulated contaminants, depending on the number of people served by the system, the source of the water supply, and the contaminants likely to be found.
- Filtration: the USEPA has criteria for determining which systems are obligated to filter water from surface water sources.
- Use of lead materials: the use of solder or flux containing more than 0.2% lead, or pipes and pipe fittings containing more than 8% lead is prohibited in public water supply systems. Public notification is required where lead is used in construction materials of the public

water supply system, or where water is sufficiently corrosive to cause leaching of lead from the distribution system or lines.

- Wellhead protection: the 1986 amendments require all states to develop Wellhead Protection Programs. These programs are designed to protect public water supplies from sources of contamination.

The USEPA developed regulations to meet the requirements of the SDWA, which are called the National Drinking Water Regulations. Found in CFR 40, these regulations are subdivided into *primary drinking water standards,* which specify maximum contaminant levels (MCLs) based on health-related criteria; and *secondary drinking water standards,* which are unenforceable "guidelines" based on both aesthetic qualities such as taste, odor, and color of drinking water, as well as nonaesthetic qualities such as corrosivity and hardness. In setting MCLs, the USEPA is required to balance the public health benefits of the standard against what is technologically and economically feasible. In this way, MCLs are different from other set standards (such as National Ambient Air Quality Standards (NAAQS)), which must be set at levels that protect public health regardless of cost or feasibility (Masters, 1991).

Note: If monitoring the contaminant level in drinking water is not economically or technically feasible, the USEPA must specify a treatment technique that will effectively remove the contaminant from the water supply or reduce its concentration. The MCLs currently cover a number of volatile organic compounds, organic chemicals, inorganic chemicals, and radionuclides, as well as microbes and *turbidity* (cloudiness or muddiness). The MCLs are based on an assumed human consumption of two liters (roughly two quarts) of water per day.

The USEPA also creates unenforceable maximum contaminant level goals (MCLG) set at levels that present no known or anticipated health effects and include a margin of safety, regardless of technological feasibility or cost. The USEPA is also required (under SDWA) to periodically review the actual MCLs to determine whether they can be brought closer to the desired MCLGs.

Note: For noncarcinogens, MCLGs are arrived at in a three-step process. The first step is calculating the *Reference Dose* (RfD) for each specific contaminant. The RfD is an estimate of the amount of a chemical that a person can be exposed to on a daily basis that is not anticipated to cause adverse systemic health effects over the person's lifetime. A different assessment system is used for chemicals that are potential carcinogens. If toxicological evidence leads to the classification of the contaminant as a human or probable human carcinogen, the MCLG is set at zero (Boyce, 1997).

3.5.2.1 Primary Standards

Categories of primary contaminants include *organic chemicals, inorganic chemicals, microorganisms, turbidity,* and *radionuclides.* Except for some microorganisms and nitrate, water that exceeds the listed MCLs will pose no immediate threat to public health. However, all these substances must be controlled, because drinking water that exceeds the standards over long periods of time may be harmful.

(1) *Organic Chemicals.* Organic contaminants for which MCLs are being promulgated are conveniently classified using the following three groupings: *synthetic organic chemicals* (SOCs), *volatile organic chemicals* (VOCs), and *Trihalomethanes* (THMs). Table 3.1 shows a partial list of maximum allowable levels for several selected organic contaminants.

Note: As we learn more from research about the health effects of various contaminants, the number of regulated organics is likely to grow. Public drinking water supplies must be sampled and analyzed for organic chemicals at least every three years.

Synthetic organic chemicals (SOCs) are man-made and are often toxic to living organisms. SOCs are compounds used in the manufacture of a wide variety of agricultural and industrial products. This group includes primarily PCBs, carbon tetrachloride, pesticides and herbicides such as 2.4-D, aldicarb, chlordane, dioxin, xylene, phenols, and thousands of other synthetic chemicals.

Note: A 1995 study of 29 midwestern cities and towns by the Washington, D.C.-based nonprofit Environmental Working Group found pesticide residues in the drinking water in nearly all of them. In Danville, Illinois, the level of the DuPont-made weed killer cyanazine was 34 times the federal standard. In Fort Wayne, Indiana, one glass of tap water contained nine kinds of pesticides. The fact is, each year, approximately 2.6 billion pounds of pesticides are used in the United States (Lewis, 1996). These pesticides find their way into water supplies, and thus present increased risk to public health.

Volatile organic chemicals (VOCs) are synthetic chemicals that readily vaporize at room temperature. These include degreasing agents, paint thinners, glues, dyes, and some pesticides. VOCs include benzene, carbon tetrachloride, 1.1.1-trichloroethane (TCA), trichlorethylene (TCE), and vinyl chloride.

Note: In water, VOCs are particularly dangerous. VOCs are absorbed through the skin through contact with water—for example, every shower or bath. Hot water allows these chemicals to evaporate rapidly; they are

TABLE 3.1. Selected Primary Standard MCLs and MCLGs for Organic Chemicals.

Contaminant	Health Effects	MCL–MCLG (mg/L)	Sources
Aldicarb	Nervous system effects	0.003–0.001	Insecticide
Benzene	Possible cancer risk	0.005–Zero	Industrial chemicals, paints, plastics, pesticides
Carbon tetrachloride	Possible cancer risk	0.005–Zero	Cleaning agents, industrial wastes
Chlordane	Possible cancer risk	0.002–Zero	Insecticide
Endrin	Nervous system, liver, kidney effects	0.002–0.002	Insecticide
Heptachlor	Possible cancer risk	0.0004–Zero	Insecticide
Lindane	Nervous system, liver, kidney effects	0.0002–0.0002	Insecticide
Pentachlorophenol	Possible cancer risk, liver, kidney effects	0.001–Zero	Wood preservative
Styrene	Liver, nervous system effects	0.1–0.1	Plastics, rubber, drug industry
Toluene	Kidney, nervous system, liver, circulatory effects	1–1	Industrial solvent, gasoline additive, chemical manufacturing
Total trihalomethanes (TTHM)	Possible cancer risk	0.1–Zero	Chloroform, drinking water chlorination by-product
Trichloroethylene (TCE)	Possible cancer risk	0.005–Zero	Waste from disposal of dry cleaning material and manufacture of pesticides, paints, waxes; metal degreaser
Vinyl chloride	Possible cancer risk	0.002–Zero	May leach from PVC pipe
Xylene	Liver, kidney, nervous system effects	10–10	Gasoline refining by-product, paint ink, detergent

Source: USEPA, May 1994, p. 6.

harmful if inhaled. VOCs can be present in any tap water, regardless of location or water source. If tap water contains significant levels of these chemicals, they pose a health threat from skin contact, even if the water is not ingested (Ingram, 1991).

Trihalomethanes (THMs) are created in the water itself as by-products of water chlorination. Chlorine (present in essentially all U.S. tap water) combines with organic chemicals to form THMs (see Figure 3.1). They include chloroform, bromodichloromethane, dibromochloromethane, and bromoform.

Note: THMs are known carcinogens—substances that increase the risk of getting cancer—and they are present at varying levels in all public tap water.

(2) *Inorganic Chemicals.* Several inorganic substances (particularly lead, arsenic, mercury, and cadmium) are of public health importance. These inorganic contaminants and others contaminate drinking water supplies either as a result of natural processes, environmental factors, or more commonly, of human activity. Some of these are listed in Table 3.2. For most inorganics, MCLs are the same as MCLGs, but the MCLG for lead is zero.

Note: In Table 3.2, the nitrate level is set at 10 mg/L, because nitrate levels above 10 mg/L pose an immediate threat to children under one year old. Excessive levels of nitrate can react with hemoglobin in blood to produce an anemic condition known as "blue babies." Treated water is sampled and tested for inorganics at least once per year (Nathanson, 1997).

(3) *Microorganisms* (or microbiological contaminants). This group of contaminants includes bacteria, viruses, and protozoa, which can cause typhoid, cholera, and hepatitis, as well as other waterborne diseases. Bacteria are closely monitored in water supplies because they can be dangerous, and because their presence is easily detected. Because tests designed to detect individual microorganisms in water are difficult to perform, in actual practice a given water supply is not tested by individually testing for specific pathogenic microorganisms. Instead, a simpler technique is used, based on testing water for evidence of any

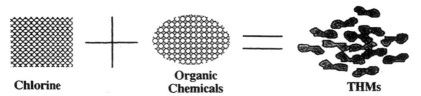

Chlorine **Organic Chemicals** **THMs**

Figure 3.1 THMs. Adapted from Ingram, C., *The Drinking Water Book,* p. 20, 1995.

TABLE 3.2. Selected Primary Standard MCLs for Inorganic Chemicals.

Contaminant	Health Effects	MCL (mg/L)	Sources
Arsenic	Nervous system effects	0.05	Geological, pesticide residues, industrial waste, smelter operations
Asbestos	Possible cancer risk	7 MFL[a]	Natural mineral deposits, A/C pipe
Barium	Circulatory system effects	2	Natural mineral deposits, paint
Cadmium	Kidney effects	0.005	Natural mineral deposits, metal finishing
Chromium	Liver, kidney, digestive system effects	0.1	Natural mineral deposits, metal finishing, textile and leather industries
Copper	Digestive system effects	TT[b]	Corrosion of household plumbing, natural deposits, wood preservatives
Cyanide	Nervous system effects	0.2	Electroplating, steel, plastics, fertilizer
Fluoride	Dental fluorosis, skeletal effects	4	Geological deposits, drinking water additive, aluminum industries
Lead	Nervous system and kidney effects, toxic to infants	TT	Corrosion of lead service lines and fixtures
Mercury	Kidney, nervous system effects	0.002	Industrial manufacturing, fungicide, natural mineral deposits
Nickel	Heart, liver effects	0.1	Electroplating, batteries, metal alloys
Nitrate	Blue-baby effect	10	Fertilizers, sewage, soil and mineral deposits
Selenium	Liver effects	0.05	Natural deposits, mining, smelting

[a]Million fibers per liter.

[b]Treatment techniques have been set for lead and copper because the occurrence of these chemicals in drinking water usually results from corrosion of plumbing materials. All systems that do not meet the action level at the tap are required to improve corrosion control treatment to reduce the levels. The action level for lead is 0.015 mg/L, and for copper it is 1.3 mg/L.

Source: USEPA, May 1994, p. 6.

fecal contamination. Coliform bacteria are used as indicator organisms whose presence suggests that the water is contaminated.

In testing for total coliforms, the number of monthly samples required is based on the population served and the size of the distribution system. Since the number of coliform bacteria excreted in feces is on the order of 50 million per gram, and the concentration of coliforms in untreated domestic wastewater is usually several million per 100 mL, that water contaminated with human wastes would have no coliforms is highly unlikely. That conclusion is the basis for the drinking water standard for microbiological contaminants, which specifies in essence that, on the average, water should contain no more than one coliform per 100 mL. The SWDA standards now require that coliforms not be found in more than five percent of the samples examined during a one month period. Known as the presence/absence concept, it replaces previous MCLs based on the number of coliforms detected in the sample.

Viruses are very common in water. If we removed a teaspoonful of water from an unpolluted lake, over one billion viruses would be present in the water (see Figure 3.2). The two most common and troublesome protozoans found in water are called *Giardia* and *Cryptosporidium* (or Crypto). In water, these protozoans occur in the form of hard-shelled cysts. Their hard covering makes them resistant to chlorination and chlorine residual that kills other organisms.

We cover microorganisms commonly found in water in much greater detail in Chapter 6.

(4) *Turbidity*. The measure of fine suspended matter in water is known as turbidity, which is mostly caused by clay, silt, organic particulates, plankton, and other microscopic organisms, ranging in size from colloidal to coarse dispersion. Turbidity in the water is measured in *nephelometric turbidity units* (NTUs); NTUs measure the amount of light scattered or reflected from the water. Officially reported in standard units or equivalent to milligrams per liter of silica of diatomaceous earth that could cause the same optical effect, turbidity testing is not required for groundwater sources.

1 billion viruses per spoonful of unpolluted lake water

Figure 3.2 Adapted from Ingram, C., *The Drinking Water Book*, p. 17, 1995.

**Radon evaporates
From an open tank**

Figure 3.3

(5) *Radionuclides*. Radioactive contamination of drinking water is a serious matter. Radionuclides (the radioactive metals and minerals that cause this contamination) come from both natural and man-made sources. Naturally occurring radioactive minerals move from underground rock strata and geologic formations into the underground streams flowing through them, and primarily affect groundwater. In water, radium-226, radium-228, radon-222, and uranium are the natural radionuclides of most concern.

Uranium is typically found in groundwater, and to a lesser degree, in some surface waters. Radium in water is found primarily in groundwater. Radon, a colorless, odorless gas and a known cancer-causing agent, is created by the natural decay of minerals. Radon is an unusual contaminant in water, because the danger arises not from drinking radon-contaminated water, but from breathing the gas after it has been released into the air. Radon dissipates rapidly when exposed to air (see Figure 3.3). When present in household water, it evaporates easily into the air, where household members may inhale it. Some experts believe that the effects of radon inhalation are more dangerous than those of any other environmental hazard.

Man-made radionuclides (more than 200 are known) are believed to be potential drinking water contaminants. Man-made sources of radioactive minerals in water are nuclear power plants, nuclear weapons facilities, radioactive materials disposal sites, and docks for nuclear-powered ships.

TABLE 3.3. National Secondary Drinking Water Standards.

Contaminants	Suggested Levels	Contaminant Effects
Aluminum	0.05–0.2 mg/L	Discoloration of water
Chloride	250 mg/L	Salty taste; corrosion of pipes
Color	15 color units	Visible tint
Copper	1.0 mg/L	Metallic taste; blue-green staining of porcelain
Corrosivity	Noncorrosive	Metallic taste; fixture staining, corroded pipes (corrosive water can leach pipe materials, such as lead, into drinking water)
Fluoride	2.0 mg/L	Dental fluorosis (a brownish discoloration of the teeth)
Foaming agents	0.5 mg/L	Aesthetic: frothy, cloudy, bitter taste, odor
Iron	0.3 mg/L	Bitter metallic taste; staining of laundry, rusty color, sediment
Manganese	0.05 mg/L	Taste; staining of laundry, black to brown color, black staining
Odor	3 threshold odor	Rotten egg, musty, or chemical smell
pH	6.5–8.5	Low pH: bitter metallic taste, corrosion; high pH: slippery feel, soda taste, deposits
Silver	0.1 mg/L	Argyria (discoloration of skin), graying of eyes
Sulfate	250 mg/L	Salty taste; laxative effects
Total Dissolved Solids (TDS)	500 mg/L	Taste and possible relation between low hardness and cardiovascular disease; also an indicator of corrosivity (related to lead levels in water); can damage plumbing and limit effectiveness of soaps and detergents
Zinc	5 mg/L	Metallic taste

Source: USEPA, May 1994, p. 6.

3.5.2.2 Secondary Standards

Secondary drinking water standards pertain to the effect drinking water has on human aesthetic preferences related to public acceptance of drinking water. They are unenforceable. A range of concentrations is established for substances that affect water only aesthetically, and have no direct effect on public health. We present secondary standards in Table 3.3.

3.6 1996 AMENDMENTS TO SDWA

After more than three years of effort, the Safe Drinking Water Act Reauthorization (one of the most significant pieces of environmental legislation passed to date) was adopted by Congress and signed into law by President William Jefferson Clinton on August 6, 1996. The new stream-

lined version of the original SDWA gives states greater flexibility in identifying and considering the likelihood for contamination in potable water
supplies, and in establishing monitoring criteria. It establishes increased
reliance on "sound science," paired with more consumer information presented in readily understandable form, and calls for increased attention to
assessment and protection of source waters.

The significance of the 1996 SDWA amendments lies in the fact that they
are a radical rewrite of the law the USEPA, states, and water systems have
been trying to implement for the past 10 years. In contrast to the 1986
amendments (which were crafted with little substantive input from the
regulated community and embraced a command-and-control approach with
compliance costs rooted in water rates), the 1996 amendments were developed with significant contributions from water suppliers and state and local
officials and embody a partnership approach that includes major new infu-
sions of federal funds to help water utilities—especially the thousands of
smaller systems—comply with the law.

In Table 3.4, we provide a summary of many of the major provisions of
the new amendments, which are as complex as they are comprehensive.

TABLE 3.4. Summary of Major Amendment Provisions for
the 1996 SDWA Regulations.

Definition	• *Constructed conveyances* such as cement ditches used primarily to supply substandard drinking water to farm workers now SDWA protected.
Contaminant Regulation	• Old contamination selection requirement (EPA regulates 25 new contaminants every three years) deleted.
	• EPA must evaluate at least five contaminants for regulation every five years, addressing the most risky first, and considering vulnerable populations.
	• EPA must issue cryptosporidium rule (enhanced surface water treatment rule) and disinfection by-product rules under agreed deadlines. The Senate provision giving industry veto power over EPA's expediting the rules was deleted.
	• EPA is authorized to address "urgent threats to health" using an expedited, streamlined process.
	• No earlier than three years after enactment, no later than the date EPA adopts the State II DBP rule, EPA must adopt a rule requiring disinfection of certain *groundwater* systems, and provide guidance on determining which systems must disinfect. USEPA may use cost-benefit provisions to establish this regulation.
Risk Assessment, Management and Communication	• Requires cost/benefit analysis, risk assessment, vulnerable population impact assessment, and development of public information materials for EPA rules.
	• Standard setting provision allows but doesn't require EPA to use risk assessment and cost/benefit analysis in setting standards.

(continued)

Standard Setting	• Cuts back Senate's process from three to two steps to issue standards, deleting requirement of Advanced Notice of Proposed Rule Making.
	• Risks to vulnerable populations must be considered.
	• Has cost/benefit and risk/risk as discretionary USEPA authority. "Sound Science" provision is limited to standard setting and scientific decisions.
	• Standard reevaluated every six years instead of every three years as current law.
Treatment Technologies for Small Systems	• Establishes new guidelines for EPA to identify best treatment technology for meeting specific regulations.
	• For each new regulation, USEPA must identify affordable treatment technologies that achieve compliance for three categories of small systems: those serving 3301–10,000, those serving 501–3300 and those serving 500 or fewer.
	• For all contaminants other than microbials and their indicators, the technologies can include package systems as well as point-of-use and point-of-entry units owned and maintained by water systems.
	• EPA has two years to list such technologies for current regulations, and one year to list such technologies for the surface water treatment rule.
	• EPA must identify best treatment technologies for the same system categories for use under variances. Such technologies do not have to achieve compliance but must achieve maximum reduction, be affordable, protect public health.
	• EPA has two years to identify variance technologies for current regulations.
Limited Alternative to Filtration	• Allows systems with fully controlled pristine watersheds to avoid filtration if EPA and State agree health is protected through other effective inactivation of microbial contaminants.
	• EPA has four years to regulate recycling of filter backwash.
Effective Date of Rules	• Extends compliance time from 18 months (current law) to three years, with available extensions of up to 5 years total.
Arsenic, Sulfate, Radon	• *Arsenic:* requires EPA to set new standard by 2001 using new standard setting language, after more research and consultation with the NAS.[a] The law authorizes $2.5 million/year for four years for research.
	• *Sulfate:* EPA has 30 months to complete a joint study with the Federal Centers for Disease Control (CDC) to establish a reliable dose-response relationship. Must consider sulfate for regulation within five years. If EPA decides to regulate sulfate, it must include public notice requirements and allow alternative supplies to be provided to at-risk populations.

Arsenic, Sulfate, Radon (continued)	• *Radon:* requires EPA to withdraw its proposed radon standard and to set a new standard in four years, after NAS conducts a risk assessment and a study of risk-reduction benefits associated with various mitigation measures. Authorizes cost/benefit analysis for radon, taking into account costs and benefits of indoor air radon control measures. States or water systems obtaining EPA approval of a multimedia radon program in accordance with EPA guidelines would only have to comply with a weaker "alternative Maximum Contaminant Level" for radon that would be based on the contribution of outdoor radon to indoor air.
State Primacy	• Primacy states have two years to adopt new or revised regulations no less stringent than federal ones and allows two or more years if EPA finds necessary and justified.
	• Provides states with interim enforcement authority between the time they submit their regulations to EPA and EPA approval.
Enforcement and Judicial Review	• Streamlines EPA administrative enforcement, increases civil penalties, clarifies enforceability of lead ban and other previously ambiguous requirements, allows enforcement to be suspended in some cases to encourage system consolidation or restructuring, requires states to have administrative penalty authority, and clarifies provisions for judicial review of final EPA actions.
Public Right to Know	"Consumer Confidence Reports" provision requires consumer to be told at least annually:
	• the levels of regulated contaminants detected in tap water
	• what the enforceable maximum contaminant levels and the health goals are for the contaminants (and what those levels mean)
	• the levels found of unregulated contaminants required to be monitored
	• information on the health effects of regulated contaminants found at levels above enforceable standards, and on health effects of up to three regulated contaminants found at levels below EPA enforceable health standards where health concerns may still exist
	• EPA's toll-free hotline for further information

(continued)

Public Right to Know (continued)	Governors can waive the requirement to mail these reports for systems serving under 10,000 people, but systems must still publish the report in the paper.
	Systems serving 500 or fewer people need only prepare the report and tell their customers it's available.
	States can later modify the content and form of the reporting requirements.
	• The public information provision modestly improves public notice requirements for violations (such as requiring "prominent" newspaper publication instead of buried classified ads). States and USEPA must prepare annual reports summarizing violations.
Variances and Exceptions	• Provisions for small system variances make minor changes to current provisions regarding exemption criteria and schedules.
	• States are authorized to grant variances to systems serving 3300 or fewer people but need EPA approval to grant variances to systems serving between 3301 and 10,000 people. Such variances are available only if EPA identifies an applicable variance technology and systems install it. ·
	• Variances only granted to systems that cannot afford to comply (as defined by state criteria that meet EPA guidelines) through treatment, alternative sources or restructuring, and when states determine that the terms of the variance ensure adequate health protection. Systems granted such variances have three years to comply with its terms and may be granted an extra two years if necessary, and states must review eligibility of such variances every five years thereafter.
	• Variances not allowed for regulations adopted prior to 1986 for microbial contaminants or their indicators.
	• EPA has two years to adopt regulations specifying procedures for granting or denying such variances and for informing consumers of proposed variances and pertinent public hearings. They also must describe proper operation of variance technologies and eligibility criteria. USEPA and the Federal Rural Utilities Service have 18 months to provide guidance to help states define affordability criteria.
	• EPA must periodically review state small system variance programs and may object to proposed variances and overturn issued variances if objections are not addressed. Also, customers of a system for which a variance is proposed can petition USEPA to object.
	• New York may extend deadlines for certain small, unfiltered systems in nine counties to comply with federal filtration requirements.

Capacity Development	• States must acquire authority to ensure that community and nontransient-noncommunity systems beginning operation after 10/1/1999, have technical, managerial, and financial capacity to comply with SDWA regulations. States that fail to acquire authority lose 20% of their annual state revolving loan fund grants.
	• States have one year to send USEPA a list of systems with a history of significant noncompliance and five years to report on the success of enforcement mechanisms and initial capacity development efforts. State primacy agencies must also provide progress reports to governors and the public.
	• States have four years to implement strategy to help systems acquire and maintain capacity before losing portions of their SRLF grants.
	• USEPA must review existing capacity programs and publish information within 18 months to help states and water systems implement such programs. USEPA has two years to provide guidance for ensuring capacity of new systems and must describe likely effects of each new regulation on capacity.
	• The law authorizes $26 million over seven years for grants to establish small water systems technology assistance centers to provide training and technical assistance. The law also authorizes $1.5 million/year through 2003 for USEPA to establish programs to provide technical assistance aimed at helping small systems achieve and maintain compliance.
Operator Certification	• Requires all operators of community and non-transient noncommunity systems be certified. EPA has 30 months to provide guidance specifying minimum standards for certifying water system operators, and states must implement a certification program within two years or lose 20% of the SRLF grants.
	• States with such programs can continue to use them as long as EPA determines they are substantially equivalent to its program guidelines.
	• EPA must reimburse states for the cost of certification training for operators of systems serving 3300 or fewer people, and the law authorizes $30 million/year through 2003 for such assistance grants
State Supervision Program	• Authorizes $100 million/year through 2003 for public water system supervision grants to states.
	• Allows EPA to reserve a state's grant should EPA assume primacy, and if needed, use SRLF resources to cover any shortfalls in TWSS appropriations.

(continued)

49

TABLE 3.4 (continued). Summary of Major Amendment Provisions for the 1996 SDWA Regulations.

Drinking Water Research	• EPA is authorized to conduct drinking water and groundwater research, and is required to develop a strategic research plan, and to review the quality of all such research.
Water Return Flows	• Repeals the provision in current law that allows businesses to withdraw water from a public water system (such as for industrial cooling purposes), then to return the used water—perhaps with contamination—to the water system's pipe.
Enforcement	• Expands and clarifies EPA's enforcement authority in primacy and nonprimacy states and provides for public hearings regarding civil penalties ranging from $5,000–$25,000. • Provides enforcement relief to systems that submit a plan to address problems by consolidating facilities or management, or transferring ownership. • States must obtain authority to issue administrative penalties, which cannot be less than $1000/day for systems serving over 10,000 people. • EPA can assess civil penalties as high as $15,000/day under its emergency powers authority.

[a]National Academy of Sciences.

3.7 IMPLEMENTING SDWA

On December 3, 1998 at the oceanfront of Fort Adams State Park, Newport, Rhode Island, in remarks by President Clinton to the community of Newport, a significant part of the 1996 SDWA and amendments were announced—the expectation being that the new requirements will protect most of the nation from dangerous contaminants, while adding only about two dollars to many monthly water bills.

The new rules, which go into effect in the year 2001, require 13,000 municipal water suppliers to use better filtering systems to screen out *Cryptosporidium* and other microbes, ensuring that U.S. community water supplies are safe from microbial contamination.

In his speech, President Clinton said:

This past summer I announced a new rule requiring utilities across the country to provide their customers regular reports on the quality of their drinking water. When it comes to the water our children drink, Americans cannot be too vigilant.

Today I want to announce three other actions I am taking. First, we're escalating our attack on the invisible microbes that sometimes creep into the water supply. . . .

Today, the new standards we put in place will significantly reduce the risk from *Cryptosporidium* and other microbes, to ensure that no community ever has to endure an outbreak like the one Milwaukee suffered.

Second, we are taking steps to ensure that when we treat our water, we do it as safely as possible. One of the great health advances to the 20th century is the control of typhoid, cholera, and other diseases with disinfectants. Most of the children in this audience have never heard of typhoid and cholera, but their grandparents cowered in fear of it, and their great-grandparents took it as a fact of life that it would take away significant numbers of the young people of their generation. But as with so many advances, there are trade-offs. We now see that some of the disinfectants we use to protect our water can actually combine with natural substances to create harmful compounds. So today I'm announcing standards to significantly reduce our exposure to these harmful byproducts, to give our families greater peace of mind with their water.

The third thing we are doing today is to help communities meet higher standards, releasing almost $800 million to help communities in all 50 states to upgrade their drinking water systems . . . to give 140 million Americans safer drinking water.

3.8 REFERENCES

AWWA. *Water Quality*, 2nd Edition. Denver, Colorado: American Water Works Association, 1995.

Bangs, R. and Kallen, C., *Rivergods*, pp. 2–3, 1985.

Boyce, A., *Introduction to Environmental Technology*. New York: Van Nostrand Reinhold, 1997.

Clark, W. C., in a speech given at Racine, Wisconsin, April 1988.

Clinton, William Jefferson, in a speech given at Newport, Rhode Island, December 3, 1998.

Ingram, C., *The Drinking Water Book*. Berkeley, California: Ten Speed Press, 1991.

Lewis, S. A., *The Sierra Club Guide to Safe Drinking Water*. San Francisco: Sierra Club Books, 1996.

Masters, G. M., *Introduction to Environmental Engineering and Science*. Englewood Cliffs, NJ: Prentice Hall, 1991.

Morrison, A., "In Third World Villages, a Simple Handpump Saves Lives." *Civil Engineering*/ASCE, pp. 68–72, October, 1983.

Nathanson, J. A., *Basic Environmental Technology: Water Supply, Waste Management, and Pollution Control*. Upper Saddle River, NJ: Prentice Hall, 1997.

Norse, E. A., in R. J. Hoage, ed., *Animal Extinctions*, pp. 3, 8, 1985.

Sierra Club, "Clean Water Timeline." In *The Planet*, Vol. 4. San Francisco: Sierra Club, October 1997.

Spellman, F. R., *Stream Ecology and Self-Purification: An Introduction for Wastewater and Water Specialists*. Lancaster, PA: Technomic Publishing Company, Inc., 1996.

USEPA, *Selected Primary Standard MCSs & MCLGs for Organic Chemicals*. Washington, D.C., USEPA 810-F-94-002, 1994.

USEPA, *Clean Water Act*. www.epa.gov, 5/23/1996.

CHAPTER 4

Drinking Water Supplies

Between earth and earth's atmosphere, the amount of water remains constant; there is never a drop more, never a drop less. This is a story of circular infinity, of a planet birthing itself. (Linda Hogan, Northern Lights, Autumn 1990)

All the water that will ever be is, right now. (National Geographic, October 1993)

4.1 INTRODUCTION

WHERE do we get our drinking water from? From what water source is our drinking water provided?

To answer those questions, we would most likely turn to one of two possibilities: our public water is provided by either a groundwater *or* surface water source—because these two sources are, indeed, the primary sources of most water supplies.

From our earlier discussion of the hydrologic or water cycle, we know that from whichever of the two sources we obtain our drinking water, the source is constantly being replenished (we hope) with a supply of fresh water. This water cycle phenomenon was best summed up by Heraclitus of Ephesus, who said, "You could not step twice into the same rivers; for other waters are ever flowing on to you."

In this chapter, we discuss one of the drinking water practitioner's primary duties—to find and secure a source of potable water for human use.

4.2 WATER SOURCES[4]

In the real estate business, location is everything. We say the same when

---

[4]Much of this section is adapted from F. R. Spellman's *The Science of Water*. Lancaster, PA: Technomic Publishing Company, Inc., pp. 18–22, 1998.

it comes to sources of water. In fact, the presence of water defines "location" for communities. Although communities differ widely in character and size, all have the common concerns of finding water for industrial, commercial, and residential use. Freshwater sources that can provide stable and plentiful supplies for a community don't always occur where we wish. Simply put, on land, the availability of a regular supply of potable water is the most important factor affecting the presence—or absence—of many life-forms. A map of the world immediately shows us that surface waters are not uniformly distributed over the Earth's surface. U.S. land holds rivers, lakes, and streams on only about four percent of its surface. The heaviest populations of any life-forms, including humans, are found in regions of the U.S. (and the rest of the world) where potable water is readily available, because lands barren of water simply won't support large populations.

Note: The volume of freshwater sources depends on geographic, landscape, and temporal variations, and on the impact of human activities.

4.2.1 JUST HOW READILY AVAILABLE IS POTABLE WATER?

Approximately 326 million cubic miles of water comprise Earth's entire water supply. Of this massive amount of water, though providing us indirectly with freshwater through evaporation from the oceans, only about three percent is fresh. Even most of the minute percentage of freshwater Earth holds is locked up in polar ice caps and in glaciers. The rest is held in lakes, in flows through soil, and in river and stream systems. Only 0.027% of Earth's freshwater is available for human consumption (see Table 4.1 for the distribution percentages of Earth's water supply).

TABLE 4.1. World Water Distribution.

Location	Percent of Total
Land areas	
Freshwater lakes	0.009
Saline lakes and inland seas	0.008
Rivers (average instantaneous volume)	0.0001
Soil moisture	0.005
Groundwater (above depth of 4000 m)	0.61
Ice caps and glaciers	2.14
Total: land areas	2.8
Atmosphere (water vapor)	0.001
Oceans	97.3
Total all locations (rounded)	100

Source: Adapted from Peavy et al., p. 12, 1985.

TABLE 4.2. Water Balance in the U.S. (in bgd).

Precipitation	4250
Evaporation and transpiration	3000
Runoff	1250
Withdrawal	310
Irrigation	142
Industry (principally utility cooling)	142
Municipal	26
Consumed (principially irrigation loss)	90
Returned to streams	220

Source: National Academy of Sciences, 1962.

We see from Table 4.1 that the major sources of drinking water are from surface water, groundwater, and from groundwater under the direct influence of surface water (i.e., springs or shallow wells).

4.2.1.1 Surface Water Supplies

Most surface water originates directly from precipitation—rainfall or snow. To gain an appreciation for the impact of this runoff on surface water supplies, let's take a look at the water balance in the United States.

Over the U.S. mainland, rainfall averages about 4,250 billion gallons per day. Of this massive amount, about 66% returns to the atmosphere through evaporation directly from the surface of lakes and rivers and transpiration from plants. This leaves about 1250 billion gallons per day to flow across or through the Earth to return to the sea (see Table 4.2). While municipal usage of water is only a small fraction of this great volume, the per capita consumption of water in the U.S. is rather high—about 150 gallons per person per day, probably because public water is relatively inexpensive in the U.S. In areas where water supplies are less readily available and thus more costly, per capita consumption is much lower, from both financial and conservation concerns.

The 1250 billion gallons per day of surface runoff water, exposed and open to the atmosphere, result from the movement of water on and just beneath the Earth's surface (overland flow). Simply put, overland flow and surface runoff are the same—water flow that has not yet reached a definite stream channel. This occurs when the rate of precipitation exceeds either the rate of interception and evapotranspiration or the amount of rainfall readily absorbed by the Earth's surface. The total land area that contributes runoff to a stream or river is called a *watershed, drainage basin,* or *catchment area.*

Specific sources of surface water include:

- rivers
- streams

- lakes
- impoundments (man-made lakes made by damming a river or stream)
- very shallow wells that receive input via precipitation
- springs affected by precipitation (flow or quantity directly dependent upon precipitation)
- rain catchments (drainage basins)
- tundra ponds or muskegs (peat bogs)

Surface water has advantages as a source of potable water. Surface water sources are usually easy to locate. Unlike groundwater, finding surface water does not take a geologist or hydrologist. Normally, surface water is not tainted with minerals precipitated from the Earth's strata.

Ease of discovery aside, surface water also presents some disadvantages: Surface water sources are easily contaminated (polluted) with microorganisms that can cause waterborne diseases and from chemicals that enter from surrounding runoff and upstream discharges. Water rights can also present problems.

As we have said, most surface water is the result of surface runoff. The amount and flow rate of this surface water is highly variable, which comes into play for two main reasons: (1) human interferences (influences) and (2) natural conditions. In some cases, surface water runs quickly off land surfaces. From a water resources standpoint, this is generally undesirable, because quick runoff does not provide enough time for the water to infiltrate the ground and recharge groundwater aquifers. Surface water that quickly runs off land also causes erosion and flooding problems. Probably the only good thing that can be said about surface water that runs off quickly is that it usually does not have enough contact time to increase in mineral content. Slow surface water off land has all the opposite effects.

Drainage basins collect surface water and direct it on its gravitationally influenced path to the ocean. The drainage basin is normally characterized as an area measured in square miles, acres, or sections. Obviously, if a community is drawing water from a surface water source, the size of its drainage basin is an important consideration.

Surface water runoff, like the flow of electricity, flows or follows the path of least resistance. Surface water within the drainage basin normally flows toward one primary watercourse (river, stream, brook, creek, etc.), unless some man-made distribution system (canal or pipeline) diverts the flow.

Surface water run off from land surfaces depends on several factors (see Figure 4.1), which include:

- *Rainfall duration*: even a light, gentle rain, if it lasts long enough, can, with time, saturate soil and allow runoff to take place.
- *Rainfall intensity*: with increases in intensity, the surface of the soil

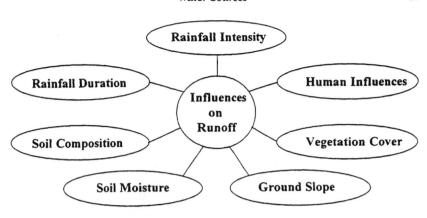

Figure 4.1 Influence on runoff.

quickly becomes saturated. This saturated soil can hold no more water; as more rain falls and water builds up on the surface, it creates surface runoff.

- *Soil moisture*: the amount of existing moisture in the soil has a definite impact on surface runoff. Soil already wet or saturated from a previous rain causes surface runoff to occur sooner than if the soil were dry. Surface runoff from frozen soil can be up to 100% of snow melt or rain runoff because frozen ground is basically impervious.
- *Soil composition:* the composition of the surface soil directly affects the amount of runoff. For example, hard rock surfaces obviously result in 100% runoff. Clay soils have very small void spaces that swell when wet; the void spaces close and do not allow infiltration. Coarse sand possesses large void spaces that allow easy flow through of water, which produces the opposite effect, even in a torrential downpour.
- *Vegetation cover:* groundcover limits runoff. Roots of vegetation and pine needles, pine cones, leaves, and branches create a porous layer (a sheet of decaying natural organic substances) above the soil. This porous "organic" sheet readily allows water into the soil. Vegetation and organic waste also act as cover to protect the soil from hard, driving rains, which can compact bare soils, close off void spaces, and increase runoff. Vegetation and groundcover work to maintain the soil's infiltration and water-holding capacity, and also work to reduce soil moisture evaporation.
- *Ground slope:* when rain falls on steeply sloping ground, up to 80% or more may become surface runoff. Gravity moves the water down the surface more quickly than it can infiltrate the surface. Water flow off flat land is usually slow enough to provide opportunity for a higher percentage of the rainwater to infiltrate the ground.

- *Human influences:* various human activities have a definite impact on surface water runoff. Most human activities tend to increase the rate of water flow. For example, canals and ditches are usually constructed to provide steady flow, and agricultural activities generally remove groundcover that would work to retard the runoff rate. On the opposite extreme, man-made dams are generally built to retard the flow of runoff.

Paved streets, tarmac, paved parking lots, and buildings are impervious to water infiltration, greatly increasing the amount of stormwater runoff from precipitation events. These man-made surfaces (which work to hasten the flow of surface water), often cause flooding to occur, sometimes with devastating consequences. In badly planned areas, even relatively light precipitation can cause local flooding. Impervious surfaces not only present flooding problems, they also do not allow water to percolate into the soil to recharge groundwater supplies—often another devastating blow to a location's water supply.

4.2.1.2 Groundwater Supply

Unbeknownst to most of us, our Earth possesses an unseen ocean. This ocean, unlike the surface oceans that cover most of the globe, is freshwater: the groundwater that lies contained in aquifers beneath Earth's crust. This gigantic water source forms a reservoir that feeds all the natural fountains and springs of Earth. But how does water travel into the aquifers that lie under Earth's surface?

Groundwater sources are replenished from a percentage of the average approximately three feet of water that falls to Earth each year on every square foot of land. Water falling to Earth as precipitation follows three courses. Some runs off directly to rivers and streams (roughly six inches of that three feet), eventually working back to the sea. Evaporation and transpiration through vegetation takes up about two feet. The remaining six inches seeps into the ground, entering and filling every interstice, each hollow and cavity. Although groundwater comprises only 1/6 of the total, (1,680,000 miles of water), if we could spread out this water over the land, it would blanket it to a depth of 1000 feet.

Almost all groundwater is in constant motion through the pores and crevices of the aquifer in which it occurs. The water table is rarely level; it generally follows the shape of the ground surface. Groundwater flows in the downhill direction of the sloping water table. The water table sometimes intersects low points of the ground, where it seeps out into springs, lakes, or streams.

Usual groundwater sources include wells and springs that are not influenced by surface water or local hydrologic events.

As a potable water source, groundwater has several advantages over surface water. Unlike surface water, groundwater is not easily contaminated. Groundwater sources are usually lower in bacteriological contamination than surface waters. Groundwater quality and quantity usually remain stable throughout the year. In the United States, groundwater is available in most locations.

As a potable water source, groundwater does present some disadvantages compared to surface water sources. Operating costs are usually higher, because groundwater supplies must be pumped to the surface. Any contamination is often hidden from view. Removing any contaminants is very difficult. Groundwater often possesses high mineral levels, and thus an increased level of hardness, because it is in contact longer with minerals. Near coastal areas, groundwater sources may be subject to saltwater intrusion.

Note: Groundwater quality is influenced by the quality of its source. Changes in source waters or degraded quality of source supplies may seriously impair the quality of the groundwater supply.

4.3 SUMMARY

Our freshwater supplies are constantly renewed through the hydrologic cycle, but the balance between the normal ratio of freshwater to salt water is not subject to our ability to change it. As our population grows and we move into lands without ready freshwater supplies, we place ecological strain upon those areas, and on their ability to support life.

Communities that build in areas without adequate local water supply are at risk in the event of emergency. Proper attention to our surface and groundwater sources, including remediation, pollution control, and water reclamation and reuse can help to ease the strain, but technology cannot fully replace adequate local freshwater supplies, whether from surface or groundwater sources.

4.4 REFERENCES

National Academy of Sciences, *National Research Council Publication 100-B*, 1962.

Peavy, H. S., et al., *Environmental Engineering*. New York: McGraw-Hill, Inc., 1985.

Spellman, F. R., *The Science of Water*. Lancaster, PA: Technomic Publishing Co., Inc., 1998.

CHAPTER 5

Drinking Water Conveyance
and Distribution

Water is the best of all things. (Pindar, circa 522 to 438 B.C., Olympian Odes)

5.1 INTRODUCTION

BEFORE we begin our discussion of drinking water conveyance and distribution systems, we must review the information we have covered to this point.

In many cases, a municipal water supply provides drinking water for use in homes and industries. This same water supply source may also be used for irrigation, for extinguishing fires, for street cleaning, for carrying wastes to treatment facilities, and for many other purposes. We stated that the two most important factors in any water supply are quality and quantity available. We now need to add a third factor to the mix: the location of the water supply relative to points of use.

Note that each type of water use has its own prerequisites. For example, food processing plants need large volumes and high water quality. Waste conveyance systems, on the other hand, require only volume.

A typical water supply system consists of six functional elements: a source or sources of supply; storage facilities (impoundment reservoirs, for example); transmission facilities used for transporting water from the point of storage to the treatment plant; treatment facilities for altering water quality; transmission and storage facilities for transporting water to intermediate points (such as water towers or standpipes); and distribution facilities for bringing water to individual users (see Figure 5.1).

Recall that when precipitation falls on a watershed or catchment area, it either flows as runoff above ground to streams and rivers, or soaks into the ground to reappear in springs or to where it can be drawn from wells. A water supply can come from a catchment area that may contain several thousands of acres of land, draining to streams whose flow is retained in impoundment

61

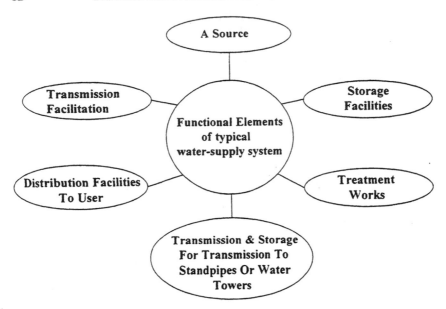

Figure 5.1 Elements of typical water supply system.

reservoirs. If a water supply is drawn from a large river or lake, the catchment area is the entire area upstream from the point of intake.

The amount of water that enters a water supply system depends on the amount of precipitation and the volume of the runoff. The annual average precipitation in the U.S. is about 30 inches, of which two-thirds is lost to the atmosphere by evaporation and transpiration. The remaining water becomes runoff into rivers and lakes, or through infiltration replenishes groundwater. Precipitation and runoff vary greatly with geography and season.

Drinking water comes from surface water and/or groundwater. Large-scale water supply systems tend to rely on surface water resources; smaller water systems tend to use groundwater.

If surface water is the source of supply for a particular drinking water supply system, the water is obtained from lakes, streams, rivers, or ponds. Storage reservoirs (artificial lakes created by constructing dams across stream valleys) can hold back higher-than-average flows and release them when greater flows are needed. Water supplies may be taken directly from reservoirs or from locations downstream of the dams. Reservoirs may serve other purposes in addition to water supply, including flood mitigation, hydroelectric power, and water-based recreation.

Groundwater is pumped from natural springs, from wells, and from infiltration galleries, basins, or cribs. Most small, and some large U.S. water systems, use groundwater as their source of supply.

Groundwater may be drawn from the pores of alluvial, glacial, or eolian deposits of granular unconsolidated material (such as sand and gravel); from the solution passages, caverns, and cleavage planes of sedimentary rocks (such as limestone, slate, and shale); and from combinations of these geologic formations. Groundwater sources may have intake or recharge areas that are miles away from points of withdrawal (water-bearing stratum or aquifer). Water quality in aquifers (geologic formations that contain water) and water produced by wells depends on the nature of the rock, sand, or soil in the aquifer where the well withdraws water. Drinking water wells may be shallow (50 feet or less) or deep (more than 1000 feet). Including the approximately 23 million Americans who use groundwater as private drinking water sources, slightly more than half of the population receives its drinking water from groundwater sources.

As the chapter title indicates, the sections that follow focus on the conveyance and distribution of drinking water. For purposes of explanation, we simplify this stage of potable water systems and treatment by postponing until Chapter 10 our discussion of the treatment process that most surface water undergoes before it is conveyed and distributed to the consumer.

5.2 SURFACE WATER/GROUNDWATER DISTRIBUTION SYSTEMS

Major water supply systems can generally be divided into two categories based on the source of water they use. The water source, in turn, impacts the design, construction, and operation of the water distribution systems. The types of systems, classified by source (see Figures 5.2 and 5.3), are:

- surface water supply systems
- groundwater supply systems

Surface water (acquired from rivers, lakes, or reservoirs) flows through an intake structure into the transmission system (Figure 5.2). For groundwater, flow moves through an intake pipe from the groundwater source, then is pumped through a transmission conduit that conveys the water to a distribution system (Figure 5.3). Groundwater is generally available in most

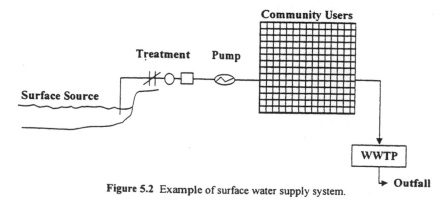

Figure 5.2 Example of surface water supply system.

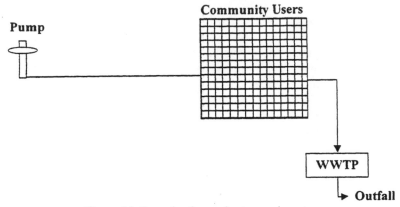

Figure 5.3 Example of groundwater supply system.

of the U.S., however, the amount available for withdrawal at any particular location is usually limited.

Surface and groundwater water supply systems may contain canals, pipes, and other conveyances; pumping plants; distribution reservoirs or tanks to assist in balancing supplies and demands for water and to control pressures; other appurtenances; and treatment works.

Note: To illustrate drinking water supply conveyance and distribution in its most basic form, we concentrate on the major components of a typical "surface" water supply and distribution system.

In a typical community water supply system, water is transported under pressure through a distribution network of buried pipes. Smaller pipes (house service lines) attached to the main water lines bring water from the distribution network to households. In many community water supply systems, pumping water up into storage tanks that store water at higher elevations than the households they serve provides water pressure. The force of gravity then "pushes" the water into homes when household taps open. Households on private supplies usually get their water from private wells. A pump brings the water out of the ground and into a small tank within the home, where the water is stored under pressure.

In cities, while the distribution system generally follows street patterns, it is also affected by topography and by the types of residential, commercial, and industrial development, as well as the location of treatment facilities and storage works. A distribution system is often divided into zones that correspond to different ground elevations and service pressures. The water pipes (mains) are generally enclosed loops, so that supply to any point can be provided from at least two directions. Street mains usually have a minimum diameter of six to eight inches to provide adequate flows for buildings and for fighting fires. The pipes connected to buildings may range down to as small as one inch for small residences.

5.2.1 SURFACE WATER INTAKE

Withdrawing water from a lake, reservoir, or river requires an intake structure. Because surface sources of water are subject to wide variations in flow, quality, and temperature, intake structures must be designed so that the required flow can be withdrawn despite these natural fluctuations. Surface water intakes consist of screened openings and conduit that conveys the flow to a sump from which it may be pumped to the treatment works. Typical intakes are towers, submerged ports, and shoreline structures. Intakes function primarily to supply the highest quality water from the source, and to protect downstream piping, equipment, and unit processes from damage or clogging as a result of floating and submerged debris, flooding, and wave action. To facilitate this, intakes should be located to consider the effects of anticipated variations in water level, navigation requirements, local currents and patterns of sediment deposition and scour, spatial and temporal variations in water quality, and the quantity of floating debris.

For lakes and impounding reservoirs where fluctuating water levels and variations in water quality with depth are common, intake structures that permit withdrawal over a wide range of elevations are typically used. Towers (Figure 5.4) are commonly used for reservoirs and lakes. A tower water intake provides ports located at several depths, avoiding the problems of water quality that stem from locating a single inlet at the bottom, since the water quality varies with both time and depth. With the exception of brief periods in spring and fall when overturns may occur, water quality is usually best close to the surface, thus intake ports located at several depths (see Figure 5.4) permit selection of the most desirable water quality in any season of the year. Submerged ports also have the advantage of remaining free from ice and floating debris. Selection of port levels must be related to characteristics of the water body (Hammer and Hammer, 1996).

Other considerations affect lake intake location, including: (1) locate as far as possible from any source of pollution; (2) factor in wind and current effects on the motion of contaminants; and (3) provide sufficient water depth (typically 20 to 30 ft.) necessary to prevent blocking of the intake by ice jams that may fill shallower lake areas to the bottom (McGhee, 1991).

River intakes are typically designed to withdraw water from slightly below the surface, to avoid both sediment in suspension at lower levels and floating debris, and if necessary, at levels low enough to meet navigation requirements. Generally, river intakes are submerged (see Figure 5.5) or screened shore intakes (see Figure 5.6). Because of low costs, the submerged type is widely used for small river and lake intakes.

Note: If they need repair, submerged intakes are not readily accessible, hampering any repair or maintenance activity. If used in lakes or reservoirs,

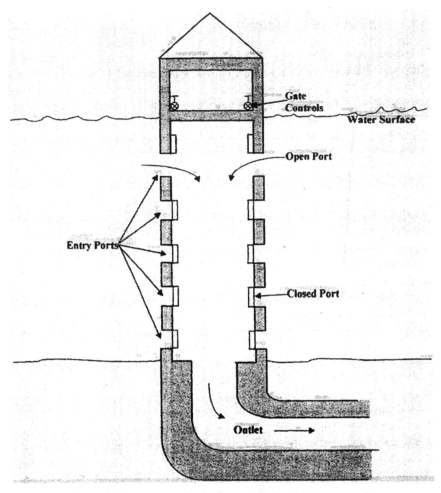

Figure 5.4 Tower water intake for a reservoir or lake water supply.

another distinct disadvantage is the lack of alternate withdrawal levels to choose the highest water quality available throughout the year.

5.2.2 SURFACE WATER DISTRIBUTION

Along with providing potable water to the household tap, water distribution systems are ordinarily designed to adequately satisfy the water requirements for a combination of domestic, commercial, industrial, and fire-fighting purposes. The system should be capable of meeting the demands placed on it at all times and at satisfactory pressures. Pipe systems, pumping stations, storage facilities, fire hydrants, house service connections,

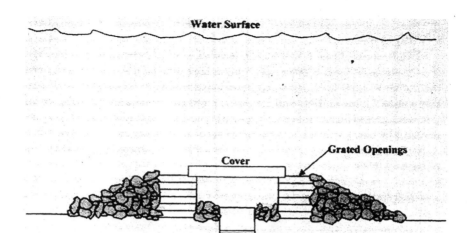

Figure 5.5 Submerged intake used for both lake and river sources.

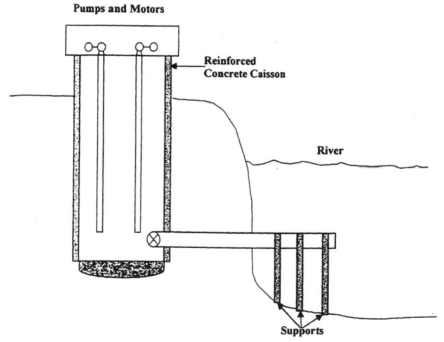

Figure 5.6 Screened shore pipe intake.

meters, and other appurtenances are the main elements of the system (Cesario, 1995).

Water is normally distributed using one of three different means: gravity distribution, pumping without storage, or pumping with storage. *Gravity distribution* is possible only when the source of supply is located substantially above the level of the community. *Pumping without storage* (the least desirable method because it provides no reserve flow and pressures fluctuate substantially) uses sophisticated control systems to match an unpredictable demand. *Pumping with storage* is the most common method of distribution (McGhee, 1991).

5.2.3 DISTRIBUTION LINE NETWORK

Distribution systems may be generally classified as grid systems, branching systems, a combination of these, or dead-end systems (see Figure 5.7). The branching system shown in Figure 5.7(a) is not the preferred distribution network, because it does not furnish supply to any point from at least two directions, and because it includes several terminals or dead ends. Normally, grid systems [see Figures 5.7(b) and 5.7(c)] are the best arrangement for distributing water. All of the arterials and secondary mains are looped and interconnected, eliminating dead ends, and permitting water circulation in such a way that a heavy discharge from one main allows drawing water from other pipes. Newly constructed distribution systems avoid the antiquated dead-end system [see Figure 5.7(d)] and such systems are often retrofitted later by incorporating proper looping.

5.2.4 SERVICE CONNECTION TO HOUSEHOLD TAP

A typical service connection consists of a pipe from the distribution system to a turnoff valve located near the property line (see Figure 5.8).

5.2.5 DISTRIBUTION STORAGE

Distribution reservoirs and other storage facilities/vessels are in place to provide a sufficient amount of water to average or equalize daily demands on the water supply system (see Figure 5.9). Storage also serves to increase operating convenience; to level out pumping requirements; to decrease power costs; to provide water during power source or pump failure; to provide large quantities of water to meet fire demands; to provide surge relief; to increase detention time; and to blend water sources.

Generally, six different types of storage are employed in storing potable water: clear wells, elevated tanks, stand pipes, ground-level reservoirs, hydropneumatic or pressure tanks, and surge tanks.

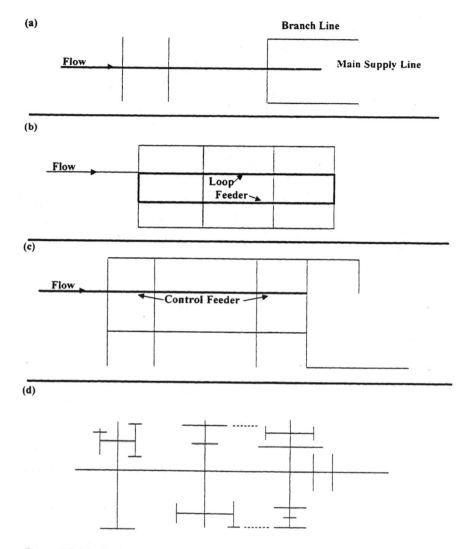

Figure 5.7 Distribution line networks: (a) branched, (b) grid, (c) combination branched/grid, (d) dead-end.

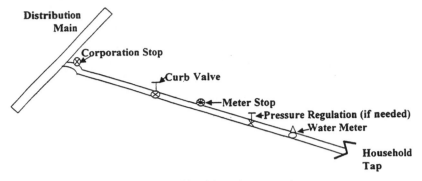

Figure 5.8 Residential service connections.

Water Demand:

    Average Daily = 400 gal/day

    Maximum Daily = (usually) 1.5 - 2 X Avg Daily

    Maximum Hourly (Peak Hourly) -- Q = 11.4 N$^{.444}$

        where
            Q = gpm
            N = # Connections

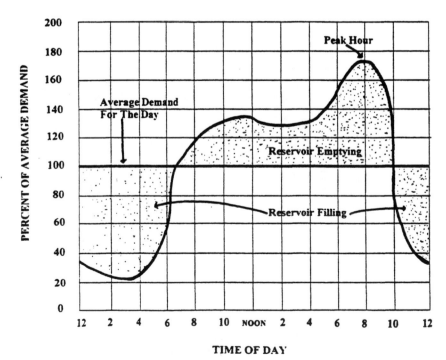

Figure 5.9 Daily variation of system demand.

(1) *Clear wells*—used to store filtered water from a treatment works, and as chlorine contact tanks (see Figure 5.10).

(2) *Elevated tanks*—located above the service zone and used primarily to maintain an adequate and fairly uniform pressure to the service zone (see Figure 5.11).

(3) *Stand pipes*—tanks that stand on the ground, with a height greater than their diameter (see Figure 5.12).

(4) *Ground-level reservoirs*—located above service area to maintain the required pressures (see Figure 5.13).

(5) *Hydropneumatic or pressure tanks*—usually used on small water systems such as with a well or booster pump. The tank is used to maintain water pressures in the system and to control the operation of the well pump or booster pump (see Figure 5.14).

(6) *Surge tanks*—not necessarily storage facilities, but used mainly to control water hammer, or to regulate water flow (see Figure 5.15).

5.2.5.1 Protective Coatings and Cathodic Protection

Storing water in a storage tank that is not properly protected and preserved from corrosion makes little sense and can be highly dangerous. Tanks in improper physical and material condition actually degrade the water stored within them. With any storage tank, any coating or preservative that will be in contact with potable water must meet the National Sanitation Foundation (NSF) Standard 61.

NSF Standard 61 lists the following types of coatings normally used in protecting tank surfaces:

- epoxy based coatings
- powdered epoxy coatings
- vitreous coatings such as glass fused to steel
- cement coatings
- polyurethane
- polymer modified asphaltic membrane
- galvanized
- lubricant
- asphaltic
- vinyl ester

The following actions or conditions may affect the life of coating systems:

- proper surface preparation prior to coating application
- quality of the coating (paint)
- workmanship

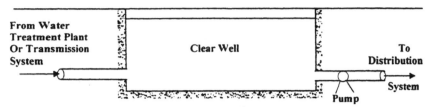

Figure 5.10 Clear well.

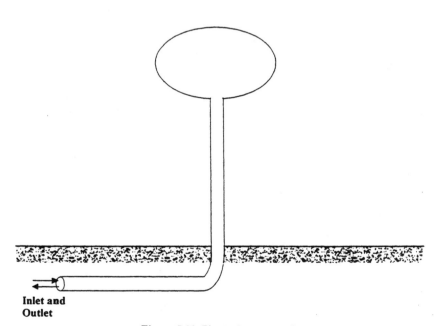

Inlet and
Outlet

Figure 5.11 Elevated storage tank.

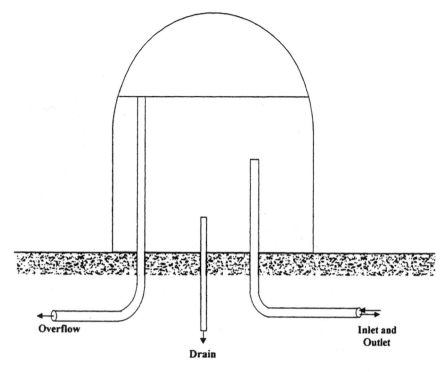

Overflow

Drain

Inlet and Outlet

Figure 5.12 Stand pipe.

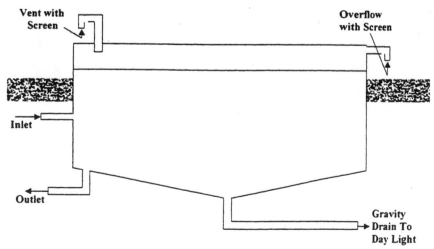

Vent with Screen

Overflow with Screen

Inlet

Outlet

Gravity Drain To Day Light

Figure 5.13 Ground-level service storage reservoir.

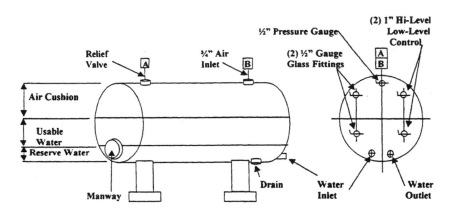

Figure 5.14 Hydropneumatic tank.

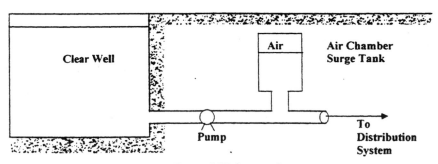

Figure 5.15 Surge tank.

- drying and aging of coating
- proper maintenance through periodic inspection; spot, partial, or complete removal of old paint; and repainting as necessary

Cathodic protection (an electrical system) is often used in preserving the integrity and material condition of potable water storage tanks, by preventing the corrosion and pitting of steel and iron surfaces in contact with water. By passing a low-voltage current through a liquid or soil in contact with the metal in such a manner that the external electromotive force renders that metal structure cathodic, corrosion is transferred to auxiliary sacrificial anodic parts. Used to corrode instead of the water storage facility corroding, the sacrificial anodes are typically made of magnesium or zinc, and must be replaced periodically as they are used up. Typically the anodes are suspended from the tank roof in warm climates, or are submerged in cold climates. The life of the sacrificial anode is about 10 years. Note that cathodic protection is not a substitute for the use of a proper interior coating system.

5.2.5.2 Water Quality Monitoring at Storage Facilities

Water stored in potable water storage facilities must be routinely properly monitored. Monitoring includes determining chlorine residual levels, turbidity, color, coliform analysis, decimal-dilution Most Probable Number (MPN) analysis, and taste and odor analysis.

5.2.5.3 Water Quality Problems in Storage/Distribution Systems

The potable water practitioner may be called upon to troubleshoot water quality problems while water is in storage facilities and distribution systems. Typical water quality problems include taste and odor, turbidity, color, and coliform present (see Table 5.1 for the types of problems commonly encountered, possible cause(s), and potential solution(s)).

5.2.5.4 Inspection of Storage Facilities

Drinking water practitioners must incorporate an annual inspection protocol for storage facilities, to verify the physical condition of the tank and to verify that the tank is maintained in sanitary condition. Recommended items to be inspected include:

(1) Tank lot
- verify access to tank location is locked if possible
- verify condition of fence
- verify surface water diverted around tank

TABLE 5.1. Troubleshooting Water Quality Problems in Storage/Distribution Systems.

Problem	Possible Cause	Potential Solution
Taste and Odor	High chlorine residual	Use breakpoint chlorination or lower chlorine dosage
	Biological (algae) growth or microorganisms	Chlorinate
	Dead end in main or tank	Flushing or eliminate dead end
Turbidity	Silt or clay in suspension, calcium carbonate, aluminum hydrate (alum), precipitated iron, oxide, microscopic floc carryover	Flushing of mains or proper operation of water treatment plant processes
Color	Decay of vegetable matter	Chlorination
	Microscopic organisms (regrowth in tank)	Chlorination
Coliform present	Contaminated distribution system	Locate and remove source
	Cross-connection	Install backflow prevention or air gap devices, flush and temporarily increase chlorine dosage
	Negative pressure in main	Repair main, increase chlorine feed rate, flush system and resample for coliform analysis. Adjust water level in tank to maintain 20 psi to all connections under all conditions of flow
	No or improper disinfection of new or repaired wells, reservoirs, or mains	Use proper disinfection procedures

- verify no problem with erosion around tank foundation, drain, and overflow
- verify condition of tank lot upkeep

(2) Water quality protection
  - exterior and interior steel/concrete watertight
  - vent shielded and screened against animals and rain
  - drain protected and screened to prevent access by animals and surface water
  - overflow protected and screened to prevent access by animals and surface water
  - roof hatch watertight
  - sidewall access watertight
  - accesses locked and/or bolted
  - all other tank openings curbed, sleeved, and covered to prevent access by animals, surface water, or rain

(3) Tank operational controls
- verify tank level indicator operable and accurate
- verify water level controls such as altitude valve and other valves
- telemetry system operational
- verify tank water level recorder operational

(4) Coating systems and corrosion control (interior and exterior)
- inspect and note cracks and peeling of coating systems
- inspect and note location of rust on metal tanks
- inspect and note pits in tank metal
- inspect and note condition of cathodic protection system

Note: A detailed inspection of the coating system should be performed by qualified persons in accordance with AWWA D101-53. The annual inspection helps to determine when a detailed inspection of the coating system is needed.

5.3 SUMMARY

Whether a community water supply is taken from surface or groundwater sources, individual community member needs are similar, whether household or industrial. Consumers need consumable water, wash water, irrigation water, and waste conveyance water. While how water arrives to their tap may cause the average consumer little thought, these systems provide the mechanics of potable water provision to the customer—an essential and valuable service, as anyone who has had to transport potable water during an emergency will tell you.

Distribution and conveyance (getting the water to the consumer) is only half the job. The other half of the job (making sure the water these systems provide is safe) is a matter of biology (Chapter 6) and disinfection (Chapter 10).

5.4 REFERENCES

Cesario, L., *Modeling, Analysis, and Design of Water Distribution Systems.* Denver: American Water Works Association, 1995.

Hammer, M. J. and Hammer, M. J., Jr., *Water and Wastewater Technology*, 3rd Edition. Englewood Cliffs, NJ: Prentice Hall, 1996.

McGhee, T. J., *Water Supply and Sewerage*, 6th Edition. New York: McGraw-Hill, Inc., 1991.

CHAPTER 6

Drinking Water Parameters: Microbiological

The subject of man's relationship to his environment is one that has been uppermost in my own thoughts for many years. Contrary to the beliefs that seem often to guide our actions, man does not live apart from the world; he lives in the midst of a complex, dynamic interplay of physical, chemical, and biological forces, and between himself and this environment there are continuing, never-ending interactions.

Unfortunately, there is so much that could be said. I am afraid it is true that, since the beginning of time, man has been a most untidy animal. But in the earlier days this perhaps mattered less. When men were relatively few, their settlements were scattered; their industries undeveloped; but now pollution has becomes one of the most vital problems of our society.... (Rachel Carson, Lost Woods, 1998, p. 32)

6.1 INTRODUCTION

DRINKING water practitioners are concerned with water supply and water purification through a treatment process. In treating water (see Chapter 11), the primary concern is producing potable water that is safe to drink (free of pathogens), with no accompanying offensive characteristics—foul taste and odor. To accomplish this, the drinking water practitioner must possess a wide range of knowledge. In short, to correctly examine raw water for pathogenic microorganisms and to determine the type of treatment necessary to ensure that the quality of the end product—potable water—meets regulatory standards, as well as to accomplish all the other myriad requirements involved in drinking water processing, the drinking water practitioner must be a combination specialist/generalist.

In the next three chapters, we concentrate on the microbiological parameters (this chapter), physical parameters (Chapter 7), and chemical parameters (Chapter 8) that drinking water practitioners must know.

As a generalist, the water practitioner requires a great deal of knowledge and skill to understand the "big picture." At the same time, drinking water practitioners fine-tune their abilities to a narrow range of focus—a focus that can be zeroed in on a single target within a broad field.

If a practitioner's narrowly focused specialty is not water microbiology, he/she must at least have enough knowledge in biological science to enable full comprehension of the fundamental factors concerning microorganisms and their relationships to one another, their effect on the treatment process, and their impact on the environment, human beings, and other organisms.

The drinking water practitioner as a generalist must know the importance of microbiological parameters and what they indicate—the potential of waterborne disease. Though true that microbiological contaminants are associated with undesirable tastes and odors, or generators of treatment problems in drinking water technology (algae and fungi, for example) that are not causes of waterborne diseases (and not regulated by public health regulations), they are still important enough to the practitioner that knowledge of them is also essential. This chapter provides fundamental knowledge of water biology (microbiology) for the water practitioner (primarily for the generalist).

Microbiology is the study of organisms that are of microscopic dimensions and thus cannot be seen except with the aid of a microscope. Microbiologists are scientists who are concerned with studying the form, structure, reproduction, physiology, metabolism, and identification of microorganisms. The microorganisms they study generally include bacteria, fungi, protozoans, algae, and viruses. These tiny organisms make up a large and diverse group of free-living forms that exist either as single cells, cell bunches, or clusters. Any and all of these organisms may be found in water.

6.1.1 CLASSIFICATION OF ORGANISMS

For centuries, scientists classified the forms of life visible to the naked eye as either animal or plant. Much of the current knowledge about living things was organized by the Swedish naturalist Carolus Linnaeus in 1735.

The importance of classifying organisms cannot be overstated, for, without a classification scheme, establishing a criteria for identifying organisms and arranging similar organisms into groups would be difficult. Probably the most important reason for classifying organisms is to make things less confusing (Wistreich and Lechtman, 1980).

Linnaeus was innovative in the classification of organisms. One of his innovations is still with us today: the *binomial system of nomenclature.* Under the binomial system, all organisms are generally described by a two-word scientific name, the *genus* and *species.* Genus and species are groups that are part of a hierarchy of groups of increasing size, based on their nomenclature (taxonomy). This hierarchy is as follows: kingdom, phylum, class, order, family, genus, and species.

Using this hierarchy and Linnaeus's binomial system of nomenclature, the scientific name of any organism (as stated previously) includes both the genus and the species name. The genus name is always capitalized, while the species name begins with a lowercase letter. On occasion, when little chance for confusion exists, the genus name is abbreviated with a single capital letter. The names are always in Latin, so they are usually printed in italics or underlined. Microbe names of interest to the drinking water practitioner include:

- *Salmonella typhi*—the typhoid bacillus
- *Escherichia coli*—a coliform bacteria
- *Giardia lamblia*—a protozoan

Escherichia coli is commonly known as simply *E. coli*, while *Giardia lamblia* is usually referred to by only its genus name, *Giardia*.

The water sciences use a simplified system of microorganism classification. Classification is broken down into the kingdoms of animal, plant, and protista. As a general rule, the animal and plant kingdoms contain all the multicell organisms, and the protists contain all single-cell organisms. Along with microorganism classifications based on the animal, plant, and protista kingdoms, microorganisms can be further classified as being *eucaryotic* or *procaryotic* (see Table 6.1). A eucaryotic organism is characterized by a cellular organization that includes a well-defined nuclear membrane. A procaryotic organism is characterized by a nucleus that lacks a limiting membrane.

To provide the fundamental knowledge of microbiology the drinking water practitioner requires, in this chapter we pursue a basic but far-reaching structured approach. To ensure currency, we include a lengthy and informa-

TABLE 6.1. Simplified Classification of Microorganisms.

Kingdom	Members	Cell Classification
Animal	Rotifers Crustaceans Worms and larvae	Eucaryotic
Plant	Ferns Mosses	
Protista	Protista Protozoa Algae Fungi	
	Bacteria Lower algae forms	Procaryotic

tive discussion of waterborne protozoans such as *Giardia, Cryptosporidium,* and others, all of which have received recent media attention.

6.2 SETTING THE STAGE

Average citizens living in the United States or Europe have heard of waterborne disease-causing microorganisms, but in this modern age they probably don't give them a second thought, even though the World Health Organization (WHO, 1984) estimates that waterborne diseases account for five million deaths annually, worldwide. Modern sanitation practices have made contraction of most of the waterborne diseases (see Table 6.2) rare in the United States and Europe. For us to forget, however, that in other areas of the world, disease-causing organisms are still in our environment—and especially that part of the environment that is water—would be foolhardy and deadly (Spellman, 1997).

The bottom line: Waterborne diseases have not been eliminated by treatment, or by this century's much improved sanitary conditions, not even in the industrialized parts of the world.

When we use the word *waterborne* in the compound term *waterborne disease*, this practice may give the uninitiated the wrong impression about

TABLE 6.2. Waterborne Disease-Causing Organisms.

Microorganism	Disease
Bacterial	
Salmonella typhi	Typhoid fever
Salmonella sp.	Salmonellosis
Shigella sp.	Shigellosis
Campylobacter jejuni	Campylobacter enteritis
Yersinia enterocolitica	Yersiniosis
Escherichia coli	
Intestinal parasites	
Entamoeba histolytica	Amebic dysentery
Giardia lamblia	Giardiasis
Cryptosporidium	Cryptosporidiosis
Viral	
Norwalk agent	—
Rotavirus	—
Enterovirus	Polio
	Aseptic meningitis
	Herpangina
Hepatitis A	Infectious hepatitis
Adenoviruses	Respiratory disease
	Conjunctivitis

water and waterborne disease. Koren (1991), for example, points out that in the water environment, water is not a medium for the *growth* of pathogenic microorganisms, but instead is a means of transmission or conveyance (a conduit—hence the name *waterborne*) of the pathogen to the place where an individual inadvertently consumes it, and the outbreak of disease begins. Again, this is contrary to the view taken by the average person. This commonly mistaken view must be replaced by the facts—and the facts are summed up simply enough by this quote:

> . . . Waterborne pathogens are not at home in water. Nothing could be further from the truth. A water-filled *ambiance* is not the environment in which the pathogenic organism would choose to live, that is, if it had such a choice. The point is that microorganisms do not normally grow, reproduce, languish, and thrive in watery surroundings. Pathogenic microorganisms temporarily residing in water are simply biding their time, going with the flow, waiting for their opportunity to meet up with their unsuspecting host or hosts. To a degree, when the pathogenic microorganism finds its host or hosts, it is finally home or may have found its final resting place. (Spellman, pp. 4–5, 1997)

6.3 BACTERIA[5]

Of all the microorganisms studied in this text, bacteria are the most widely distributed, the smallest in size, the simplest in morphology (structure), the most difficult to classify, and the hardest to identify. Because of considerable diversity, even providing a descriptive definition of a bacterial organism is difficult. About the only generalization that can be made is that bacteria are single-celled plants, are procaryotic (the nucleus lacks a limiting membrane), are seldom photosynthetic, and reproduce by binary fission.

Bacteria are found everywhere in our environment—in the soil, in the air, and in water. Bacteria are also present in and on the bodies of all living creatures, including humans. Most bacteria do not cause disease; they are not all pathogenic. Many bacteria carry on useful and necessary functions related to the life of larger organisms.

When we think about bacteria in general terms, we usually think of the damage they cause. For example, Black-Covilli (1992) points out that "the form of water pollution that poses the most direct menace to human health is bacteriological contamination" (p. 23). This is partly the reason that bacteria are of great significance to water specialists. For water treatment personnel tasked with providing the public with safe, potable water, controlling and/or eliminating disease-causing bacteria pose a constant challenge (see Table 6.3).

[5]Adapted from F. R. Spellman's *Microbiology for Water/Wastewater Operators*. Lancaster, PA: Technomic Publishing Company, Inc., pp. 19–80, 1997.

TABLE 6.3. Bacterial Agents That Cause Human Intestinal
Diseases Disseminated by Drinking Water.

Microorganism	Disease
Salmonella typhi	typhoid fever
Salmonella paratyphi-A	paratyphoid fever
Salmonella	salmonellosis, enteric fever
Shigella sp.	bacillary dysentery
Vibrio cholerae	cholera
Leptospira	leptospirosis
Yersinia enterocolitica	gastroenteritis
Francisella tularensis	tularemia
Escherichia coli	gastroenteritis
Pseudomonas aeruginosa	various infections
Edwardsiella, Proteus, Serratia	gastroenteritis

The conquest of disease has placed bacteria high on the list of microorganisms of great interest to the scientific community. This interest has spawned much accomplishment toward enhancing our understanding of bacteria. However, we still have a great deal to learn about bacteria. Thomas (1982) points out that "we are still principally engaged in making observations and collecting facts, trying wherever possible to relate one set of facts to another but still lacking much of a basis for grand unifying theories" (p. 71).

One of the important "facts" about bacteria that we still lack complete understanding of is the infecting dose. Determining, for example, the number of viable pathogenic cells necessary to produce infections is difficult. The National Academy of Sciences (1977; 1982) reported values varying from 10^3–10^9 pathogenic cells per person, with subjects infected representing from 1%–95% of the total subjects tested.

Note that other factors such as age and general health, as well as previous exposure, are important. Additional significant influencing factors include the survival of an organism in water, water temperature, and the presence of colloidal matter in water.

6.3.1 BACTERIAL CELLS: SHAPES, FORMS, SIZES, AND ARRANGEMENTS

Since the nineteenth century, scientists have known that all living things, whether animal or plant, are made up of cells. The fundamental unit of all living matter, no matter how complex, is the cell. A typical cell is a single entity, isolated from other cells by a membrane or cell wall. The cell membrane contains protoplasm, the living material found within it, and the nucleus. In a typical mature plant cell, the cell wall is rigid and is composed of nonliving material, while in the typical animal cell, the wall is an elastic living membrane. Cells exist in a very great variety of sizes and shapes, as well as having a great variety of functions. Their size ranges from bacteria too small to be seen with the light microscope to the largest known single cell, the ostrich egg. Microbial cells also have an extensive size range, some being larger than human cells (Kordon, 1993).

Bacteria come in three shapes: elongated rods called *bacilli*, rounded or spherical cells called *cocci*, and spirals (helical and curved) called *spirilla* (for the less rigid form) and *spirochaete* for those that are flexible). Elongate rod-shaped bacteria may vary considerably in length; have square, round, or pointed ends; and be motile (able to move) or nonmotile. The spherical-shaped bacteria may occur singly, in pairs, in tetrads, in chains, and in irregular masses. The helical and curved spiral-shaped bacteria exist as slender spirochaetes, spirillum, and bent rods (see Figure 6.1).

Bacterial cells are usually measured in microns (μ) or micrometers (μm); 1 μm = 0.001 or 1/1,000 of a millimeter (mm). A typical rod-shaped coliform bacterial cell is about 2 μm long and about 0.7 microns wide. The size of a cell changes with time during growth and death.

Viewed under a microscope, bacterial cells may be seen as separate (individual) cells, or as cells in groupings. According to the species, cells may appear in pairs (diplo), chains, groups of four (tetrads), cubes (sarcinae), and clumps. Long chains of cocci result when cells adhere after repeated divisions in one plane; this pattern is seen in the genera *Enterococcus* and *Lactococcus*. In the genus *Sarcina*, cocci divide in three planes, producing cubical packets of eight cells (tetrads). The exact shape of rod-shaped cells varies, especially at the end of the rod. The rod's end may be flat, cigar shaped, rounded or bifurcated. While many rods do occur singly, they may remain together after division to form pairs or chains (see Figure 6.1). Frequently, these characteristic arrangements are useful in bacterial identification.

6.3.2 STRUCTURE OF THE BACTERIAL CELL

The structural form and various components of the bacterial cell is probably best understood by referring to the simplified diagram of a rod-form bacterium shown in Figure 6.2. When studying Figure 6.2, note that cells of different species may differ greatly, both in structure and chemical composition; for this reason no typical bacterium exists. Figure 6.2 shows a generalized bacterium used for the discussion that follows. Not all bacteria have all of the features shown in the figure, and some bacteria have structures not shown in the figure.

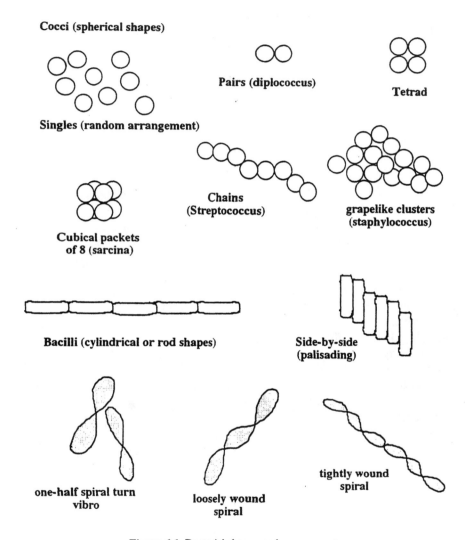

Figure 6.1 Bacterial shapes and arrangements.

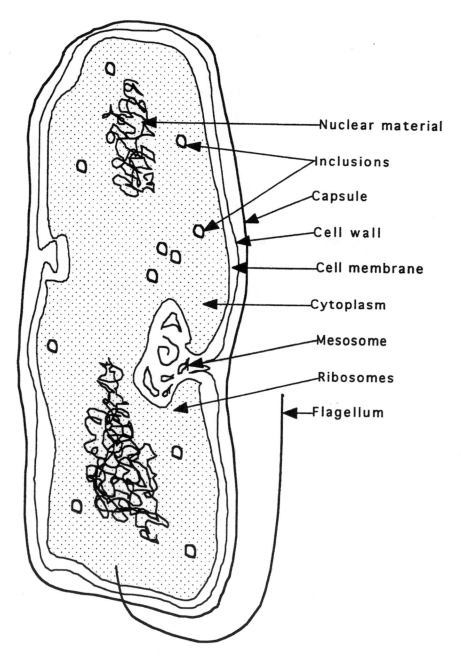

Nuclear material

Inclusions

Capsule

Cell wall

Cell membrane

Cytoplasm

Mesosome

Ribosomes

Flagellum

Figure 6.2 Bacterial cell.

6.3.2.1 Capsules

Bacterial *capsules* (see Figure 6.2) are organized accumulations of gelatinous material on cell walls, in contrast to *slime layers* (a water secretion that adheres loosely to the cell wall and commonly diffuses into the cell), which are unorganized accumulations of similar material. The capsule is usually thick enough to be seen under the ordinary light microscope (macrocapsule), while thinner capsules (microcapsules) can be detected only by electron microscopy (Singleton and Sainsbury, 1994).

The production of capsules is determined largely by genetics as well as environmental conditions, and depends on the presence or absence of capsule-degrading enzymes and other growth factors. Varying in composition, capsules are mainly composed of water; the organic contents are made up of complex polysaccharides, nitrogen-containing substances, and polypeptides.

Capsules confer several advantages when bacteria grow in their normal habitat. For example, they help to (1) prevent desiccation; (2) resist phagocytosis by host phagocytic cells; (3) prevent infection by bacteriophages; and (4) aid bacterial attachment to tissue surfaces in plant and animal hosts or to surfaces of solid objects in aquatic environments. Capsule formation often correlates with pathogenicity.

6.3.2.2 Flagella

Many bacteria are motile, and this ability to move independently is usually attributed to a special structure, the *flagella* (singular: flagellum). Depending on species, a cell may have a single flagellum (see Figure 6.2) (*monotrichous* bacteria; *trichous* means "hair"); one flagellum at each end (*amphitrichous* bacteria; *amphi* means "on both sides"); a tuft of flagella at one or both ends (*lophotrichous* bacteria; *lopho* means "tuft"); or flagella that arise all over the cell surface (*peritrichous* bacteria; *peri* means "around").

A flagellum is a threadlike appendage extending outward from the plasma membrane and cell wall. Flagella are slender, rigid, locomotor structures, about 20 μm across and up to 15 or 20 μm long.

Flagellation patterns are very useful in identifying bacteria and can be seen by light microscopy, but only after being stained with special techniques designed to increase their thickness. The detailed structure of flagella can be seen only in the electron microscope.

Bacterial cells benefit from flagella in several ways. They can increase the concentration of nutrients or decrease the concentration of toxic materials near the bacterial surfaces by causing a change in the flow rate of fluids. They can also disperse flagellated organisms to areas where colony formation can take place. The main benefit of flagella to organisms is their increased ability to flee from areas that might be harmful.

6.3.2.3 Cell Wall

The main structural component of most procaryotes is the rigid *cell wall*. Functions of the cell wall include (1) providing protection for the delicate protoplast from osmotic lysis (bursting); (2) determining a cell's shape; (3) acting as a permeability layer that excludes large molecules and various antibiotics and plays an active role in regulating the cell's intake of ions; and (4) providing a solid support for flagella.

Cell walls of different species may differ greatly in structure, thickness, and composition. The cell wall accounts for about 20–40% of a bacterium's dry weight.

6.3.2.3.1 Gram Stain

Microbial cells are nearly transparent when observed by light microscopy, and hence are difficult to see. The most common method for observing cells is by the use of stained preparations. Dyes are used to stain cells, which increases their contrast so that they can be more easily observed in the light microscope.

Simple cell staining techniques depend upon the fact that bacterial cells differ chemically from their surroundings and thus can be stained to contrast with their environment. Microbes also differ from one another chemically and physically, and therefore may react differently to a given staining procedure. This is the basic principle of *differential staining*—so named because this type of procedure does not stain all kinds of cells equally.

The Gram staining procedure was developed in the 1880s by Hans Christian Gram, a Danish bacteriologist. Gram discovered that microbes could be distinguished from surrounding tissue. Gram observed that some bacterial cells exhibit an unusual resistance to decolorization. He used this observation as the basis for a differential staining technique.

The Gram differentiation is based upon the application of a series of four chemical reagents: primary stain, mordant, decolorizer, and counterstain. The purpose of the primary stain, crystal violet, is to impart a blue or purple color to all organisms regardless of their Gram reaction. This is followed by the application of Gram's iodine, which acts as a mordant (fixer) by enhancing the union between the crystal violet stain and its substrate by forming a complex. The decolorizing solution of 95% ethanol extracts the complex from certain cells more readily than others. In the final step, a counterstain (safranin) is applied, to see those organisms previously decolorized by removal of the complex. Those organisms retaining the complex are gram-positive (blue or purple), whereas those losing the complex are gram-negative (red or pink).

6.3.2.3.1.1 GRAM-POSITIVE CELL WALLS

Normally, the thick, homogenous cell walls of gram-positive bacteria are composed primarily of a complex polymer, which often contains linear heteropolysaccharide chains bridged by peptides to form a three-dimensional netlike structure and envelop the protoplast. Gram-positive cells usually also contain large amounts of *teichoic* acids: typically, substituted polymers or ribitol phosphate and glycerol phosphate. Amino acids or sugars such as glucose are attached to the ribitol and glycerol groups.

Teichoic acids are negatively charged and help give the gram-positive cell wall its negative charge. Growth conditions can affect the composition of the cell wall; for example, the availability of phosphates affects the amount of teichoic acid in the cell wall of *Bacillus*. Teichoic acids are not present in gram-negative bacteria.

6.3.2.3.1.2 GRAM-NEGATIVE CELL WALLS

Gram-negative cell walls are much more complex than gram-positive walls. The gram-negative wall is about 20–30 μm thick and has a distinctly layered appearance under the electron microscope. The thin inner layer consists of peptidoglycan and constitutes no more than 10% of the wall weight. In *E. coli*, the gram-negative walls are about 1 μm thick and contain only one or two layers of peptidoglycan.

The outer membrane lies outside the thin peptidoglycan layer and is essentially a lipoprotein bilayer. The outer membrane and peptidoglycan are so firmly linked by this lipoprotein that they can be isolated as one unit.

6.3.2.4 Plasma Membrane (Cytoplasmic Membrane)

Surrounded externally by the cell wall and composed of a lipoprotein complex, the *plasma membrane* is the critical barrier, separating the inside from outside of the cell. About 7–8 μm thick and comprising 10–20% of a bacterium's dry weight, the plasma membrane controls the passage of all material into and out of the cell. The inner and outer faces of the plasma membrane are embedded with water-loving (hydrophilic) lips, whereas the interior is hydrophobic. Control of material into the cell is accomplished by screening, as well as by electric charge. The plasma membrane is the site of the surface charge of the bacteria.

In addition to serving as an osmotic barrier that passively regulates the passage of material into and out of the cell, the plasma membrane participates in the active transport of various substances into the bacterial cell. Inside the membrane, many highly reactive chemical groups guide the incoming material to the proper points for further reaction. This active transport system provides bacteria with certain advantages, including the ability to maintain a fairly constant intercellular ionic state in the presence of varying external ionic concentrations. In addition to participating in the uptake of nutrients, the cell membrane transport system participates in waste excretion and protein secretions.

6.3.2.5 Cytoplasm

Within a cell and bounded by the cell membrane is a complicated mixture of substances and structures called the *cytoplasm*. The cytoplasm is a water-based fluid containing ribosomes, ions, enzymes, nutrients, storage granules (under certain circumstances), waste products, and various molecules involved in synthesis, energy metabolism, and cell maintenance.

6.3.2.6 Mesosome

A common intracellular structure found in the bacterial cytoplasm is the mesosome. *Mesosomes* are invaginations of the plasma membrane in the shape of tubules, vesicles, or lamellae. Their exact function is unknown. Currently many bacteriologists believe that mesosomes are artifacts generated during the fixation of bacteria for electron microscopy (Singleton and Sainsbury, 1994).

6.3.2.7 Nucleoid (Nuclear Body or Region)

The *nuclear region* of the procaryotic cell is primitive and a striking contrast to that of the eucaryotic cell. Procaryotic cells lack a distinct nucleus, the function of the nucleus being carried out by a single, long, double-strand of deoxyribonucleic acid (DNA) that is efficiently packaged to fit within the nucleoid. The nucleoid is attached to the plasma membrane. A cell can have more than one nucleoid when cell division occurs after the genetic material has been duplicated.

6.3.2.8 Ribosomes

The bacterial cytoplasm is often packed with ribosomes. *Ribosomes* are minute, rounded bodies made of ribonucleic acid (RNA) and are loosely attached to the plasma membrane. Ribosomes are estimated to account for about 40% of a bacterium's dry weight; a single cell may have as many as 10,000 ribosomes. Ribosomes are the site of protein synthesis and are part of the translation process; they are commonly called the "powerhouses of the cell."

6.3.2.9 Inclusions

Inclusions (or storage granules) are often seen within bacterial cells. Some inclusion bodies are not bound by a membrane and lie free in the cytoplasm. Other inclusion bodies are enclosed by a single-layered membrane about 2–4 μm thick. Many bacteria produce polymers that are stored as granules in the cytoplasm.

6.3.3 CHEMICAL COMPOSITION OF A BACTERIA

Bacteria, in general, are composed primarily of water (about 80%) and of dry matter (about 20%). The dry matter consists of both organic (90%) and inorganic (10%) components. All basic elements from protoplasm must be derived from the liquid environment, and if the environment is deficient in vital elements, the cell shows a characteristic lack of development.

Note: The normal growth of a bacterial cell in excess nutrients results in a cell of definite chemical composition. The growth, however, involves a coordinated increase in the mass of its constituent parts, not solely an increase in total mass.

6.3.4 METABOLISM

Metabolism refers to the bacteria's ability to grow in any environment. The metabolic process refers to the chemical reactions that occur in living cells. In this process, *anabolism* works to build up cell components and *catabolism* breaks down or changes the cell components from one form to another.

Metabolic reactions require energy, as does locomotion and the uptake of nutrients. Many bacteria obtain their energy by processing chemicals from the environment through *chemosynthesis*. Other bacteria obtain their energy from sunlight through *photosynthesis*.

6.3.4.1 Chemosynthesis

The synthesis of organic substances such as food nutrients using the energy of chemical reactions is called *chemosynthesis*. A bacterium that obtains its carbon from carbon dioxide is called *autotrophic*. Bacteria that obtain carbon through organic compounds are called *heterotrophic* (see Figure 6.3).

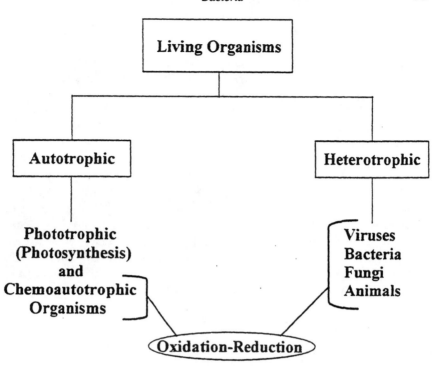

Figure 6.3 Autotrophic and heterotrophic organisms in relation to their means of obtaining energy.

6.3.4.2 Autotrophic Bacteria

Organisms that can synthesize organic molecules needed for growth from inorganic compounds using light or another source of energy are called *autotrophs*. For their carbon requirements, autotrophs are able to use (fix) carbon dioxide to form complex organic compounds.

6.3.4.3 Heterotrophic Bacteria

Most bacteria are not autotrophic. They cannot use carbon dioxide as a major source of carbon and therefore must rely upon the presence of more reduced, complex molecules (mostly derived from other organisms) for their carbon supply. Bacteria that need complex carbon compounds are called *heterotrophs*. Heterotrophs use a vast range of carbon sources, including fatty acids, alcohols, sugars, and other organic substances. Heterotrophic bacteria are widespread in nature, and include all those species that cause disease in man, other animals, and plants.

6.3.5 CLASSIFICATION

Classifying bacteria and other microbes is complicated because of the enormous variety of microorganisms that differ widely in metabolic and structural properties. Some microorganisms are plant-like, others are animal-like, and still others are totally different from all other forms of life.

As an example of the classification process, consider microorganisms in terms of their activities: Bacteria can be classified as *aerobic, anaerobic,* or *facultative*. An aerobe must have oxygen to live. At the other extreme, the same oxygen would be toxic to an anaerobe (bacteria that live without oxygen). Facultative bacteria are capable of growth under aerobic or anaerobic conditions.

Because bacteria have so many forms, their proper classification or identification requires a systematic application of procedures that grow, isolate, and identify the individual varieties. These procedures are highly specialized and technical. Ultimately, bacteria are characterized based on observation and experience. Fortunately, as Singleton (1992) points out, certain classification criteria have been established to help in the sorting process:

(1) Shape
(2) Size and structure
(3) Chemical activities
(4) Types of nutrients needed
(5) Form of energy used
(6) Physical conditions needed for growth
(7) Ability to cause disease (pathogenic or nonpathogenic)
(8) Staining behavior (Gram stain)

Using these criteria, and based on observation and experience, the bacteria can be identified from descriptions published in *Bergey's Manual of Determinative Bacteriology* (9th Edition, 1993).

6.3.6 FECAL COLIFORM BACTERIA: INDICATOR ORGANISMS

Fecal coliform are bacteria that live in the digestive tract of warm-blooded animals. They are excreted in the solid wastes of humans and other mammals. Fecal coliform generally enter the water via:

• improperly treated wastewater from municipal systems, septic systems, or combined sewer overflows
• runoff from animal stockyards, pastures, and rangeland
• inadequately captured wastes from human activities such as construction or camping

Fecal coliforms generally do not pose a danger to people or animals. Where fecal coliforms are present, however, disease-causing bacteria are usually also present. Unlike fecal coliforms, disease-causing bacteria generally do not survive outside the body of animals long enough in the water to be detected. This makes their direct monitoring difficult. Drinking water practitioners and public health officials consider the presence of fecal coliform an indicator of disease bacteria in the water.

The presence of fecal coliforms tends to affect humans more than it does aquatic creatures, though not exclusively.

- Bacteria associated with fecal coliform can cause significant disease in humans, such as diarrhea, dysentery, cholera, and typhoid fever. Some of these bacteria can also cause infection in open wounds.
- Untreated fecal material that contains fecal coliforms adds excess organic material to the water. The decay of this material depletes the water of oxygen, which may kill fish and other aquatic life.
- Reduction of fecal coliforms in wastewater may require use of chlorine and other disinfectant chemicals. Such materials may kill bacteria essential to the proper balance of the aquatic environment, endangering the survival of species dependent on those bacteria. Higher levels of fecal coliform require higher levels of chlorine, threatening those aquatic organisms.

6.3.6.1 Total Coliform Rule

The Total Coliform Rule (40 CFR 141.21) is the part of the Safe Drinking Water Act that discusses detection and removal of bacterial contamination in drinking water. The Total Coliform Rule applies to every public water system. Each public water system must take at least one coliform sample every year and submit the results of that sample to the State Drinking Water Program for compliance purposes.

6.3.6.2 Fecal Coliform Testing Procedures

Extensive research was conducted and is ongoing in the attempt to compare the presence and the significance of specific organisms in water to the traditional coliform group and waterborne diseases, with the goal of pinpointing the best indicator of contamination in water. The search continues for a quick, economic, reliable determination for possible use in routine examination, or at least during outbreaks of the waterborne diseases.

Federal regulations cite two approved methods for the determination of fecal coliform in water: (1) the Multiple Tube Fermentation Technique for members of the Coliform group and (2) the Membrane Filter Technique

(MF) for members of the Coliform group. The multiple tube method is still used in many labs, because the membrane filter method is not applicable to turbid samples. However, the membrane filter method takes less time and provides a more direct count of the coliforms than the multiple tube method does. It also requires less laboratory equipment. The bottom line? Drinking water practitioners need to understand the essential differences between these two tests. But before we explain the differences, we need to lay the groundwork by explaining testing preparations.

Note: Because the MF procedure can yield low or highly variable results for chlorinated water, the USEPA requires verification of results using the MPN (Most Probable Number) procedure to resolve any controversies.

Note: We discuss each of these procedures briefly in the following sections. However, do not attempt to perform the fecal coliform test using the summary information provided within this text. Instead, refer to the appropriate reference cited in the Federal Regulations or *Standard Methods* (20th Edition, 1999) for a complete discussion of these procedures.

Note: Both methods are still recognized as sufficiently reliable with relatively simple techniques and economic equipment to be run as often as required by the monitoring activity of water quality control.

6.3.6.2.1 Testing Preparations

Whenever microbiological testing of water samples is performed, certain general considerations and techniques are required. Since these considerations are similar for each test procedure, we review them here prior to our specific discussion of the two test methods.

- *Reagents* and *Media*—all reagents and media used in performing microbiological tests on water samples must meet the standards specified in the reference cited in Federal Regulations.
- *Reagent Grade Water*—deionized water that is tested annually and found to be free of dissolved metals and bactericidal or inhibitory compounds is preferred for use in preparing culture media and test reagents, although distilled water may be used.
- *Chemicals*—all chemicals used in fecal coliform monitoring must be ACS reagent grade or equivalent.
- *Media*—to ensure uniformity in the test procedures the use of dehydrated media is recommended. Sterilized prepared media in sealed test tubes, ampules, or in dehydrated media pads are also acceptable for use in this test.
- *Glassware and Disposable Supplies*—all glassware, equipment, and supplies used in microbiological testing should meet the standards specified in the references cited in the Federal Regulations.

All glassware used for bacteriological testing must be thoroughly cleaned, using a suitable detergent and hot water. The glassware should be rinsed with hot water to remove all traces of residual from the detergent, and finally rinsed with distilled water. Laboratories should use a detergent certified to meet bacteriological standards, or at a minimum, rinse all glassware after washing with two tap water rinses followed by five distilled water rinses.

For sterilization of equipment, the hot air sterilizer or autoclave can be used. When using the hot air sterilizer, all equipment should be wrapped in high quality (Kraft) paper or placed in containers prior to hot air sterilization. All glassware (except those in metal containers) should be sterilized for a minimum of 60 minutes at 170°C. Sterilization of glassware in metal containers should require a minimum of two hours. Hot air sterilization can't be used for liquids.

When using an autoclave, sample bottles, dilution water, culture media, and glassware may be sterilized by autoclaving at 121°C for 15 minutes.

Dilution water used for making sample serial dilutions is prepared by adding 1.25 mL of stock buffer solution and 5.0 mL of magnesium chloride solution to 1000 mL of distilled or deionized water. The stock solutions of each chemical should be prepared as outlined in the reference cited in the Federal Regulations. The dilution water is then dispensed in sufficient quantities to produce 9 or 99 mL in each dilution bottle following sterilization. If the membrane filter procedure is used, additional 60–100 mL portions of dilution water should be prepared and sterilized to provide rinse water required by the procedure.

At times, the density of the organisms in a sample makes accurately determining the actual number of organisms in the sample difficult. When this occurs, the sample size may need to be reduced to as little as one millionth of a milliliter. To obtain such small volumes, a technique known as serial dilutions has been developed.

6.3.6.3 Multiple Tube Fermentation Technique

The *multiple tube fermentation technique* (MTF) for fecal coliform testing of water, solid, or semisolid samples is based on the fact that coliform organisms can use lactose (the sugar occurring in milk) as food and produce gas in the process. For use in waste testing, the procedure normally requires use of the presumptive and confirming or completed test procedures. It is recognized as the method of choice for any samples that may be controversial (enforcement related).

When multiple tubes are used in the fermentation technique, results of the examination of replicate tubes and dilutions are reported in terms of the *Most Probable Number* (MPN) of organisms present. This number, based on certain probability formulas, is an estimate of the mean density of coliforms in the sample. Coliform density provides the best assessment of water treatment effectiveness and the sanitary quality of untreated water.

The production of gas in a single fermentation tube may indicate the presence of coliforms, but it gives no indication as to the concentration of bacteria in the sample; a coliform count cannot be obtained directly. The gas bubble created could be the result of one bacterium or millions. The precision of each test depends on the number of tubes used. The most satisfactory information will be obtained when the largest sample inoculum shows no gas at all in a majority of the tubes. Bacterial density can be estimated by the formula given, or from the table using the number of positive tubes in multiple dilutions. The number of sample portions selected is governed by the desired precision of the result. MPN tables are based on the assumption of a Poisson distribution (random dispersion). However, if the sample is not adequately shaken before the portions are removed or if clumping of bacterial cells occurs, the MPN value will underestimate the actual bacterial density.

The technique utilizes a two-step incubation procedure. The sample dilutions are first incubated in lauryl (sulfonate) tryptose broth for 24–48 hours (presumptive test). Positive samples are then transferred to EC broth and incubated for an additional 24 hours (confirming test). Positive samples from this second incubation are used to statistically determine the MPN from the appropriate reference chart.

A single media, 24-hour procedure is also acceptable. In this procedure, sample dilutions are inoculated in A-1 media and incubated for three hours at 35°C, then incubated the remaining 20 hours at 44.5°C. Positive samples from these inoculations are then used to statistically determine the MPN value from the appropriate chart.

6.3.6.3.1 Fecal Coliform MPN Presumptive Test Procedure

(1) Prepare dilutions and inoculate five fermentation tubes for each dilution.
(2) Cap all tubes and transfer to incubator.
(3) Incubate 24 + 2 hrs. at 35 ± 0.5°C.
(4) Examine tubes for gas: gas present = positive test – transfer; no gas = continue incubation.
(5) Incubate total time 48 ± 3 hrs. at 35 ± 0.5 °C.
(6) Examine tubes for gas: gas present = positive test – transfer; no gas = negative test

Note: Keep in mind that the fecal coliform MPN confirming procedure or fecal coliform procedure using A-1 broth test is used to determine the MPN/100 mL.

Note: The multiple tube fermentation MPN procedure for fecal coliform determinations requires a minimum of three dilutions with five tubes/dilution.

6.3.6.3.2 Calculation of Most Probable Number (MPN)/100 mL

The calculation of the MPN test results requires selection of a valid series of three consecutive dilutions. The number of positive tubes in each of the three selected dilution inoculations is used to determine the MPN/100 mL. In selecting the dilution inoculations to be used in the calculation, each dilution is expressed as a ratio of positive tubes per tubes inoculated in the dilution, i.e., 3 positive/5 inoculated (3/5). Several rules must be followed in determining the most valid series of dilutions. In the following examples, four dilutions were used for the test.

(1) Using the confirming test data, select the highest dilution showing all positive results (no lower dilution showing less than all positive) and the next two higher dilutions.

(2) If a series shows all negative values (with the exception of one dilution), select the series that places the only positive dilution in the middle of the selected series.

(3) If a series shows a positive result in a dilution higher than the selected series (using Rule 1), it should be incorporated into the highest dilution of the selected series.

After selecting the valid series, the MPN/100 mL is determined by locating the selected series on the MPN reference chart. If the selected dilution series matches the dilution series of the reference chart, the MPN value from the chart is the reported value for the test. If the dilution series used for the test does not match the dilution series of the chart, the test result must be calculated.

$$\text{MPN} / 100\,\text{mL} = \text{MPN}_{chart} \times \frac{\text{Sample Volume in First Dilution}_{chart}}{\text{Sample Volume in First Dilution}_{sample}} \quad (6.1)$$

Example 6.1

Using the results recorded below, calculate the MPN/100 mL of the example.

mL of Sample in Each Serial Dilution	Positive Tubes/Tubes Inoculated
10.0	5/5
1.0	5/5
0.1	3/5
0.01	1/5
0.001	1/5

Solution:

(1) Select the highest dilution (tube with the lowest amount of sample) with all positive tubes (1.0 mL dilution). Select the next two higher dilutions (0.1 mL and 0.01 mL). In this case, the selected series will be 5-3-1.

MPN Reference Chart

10	1.0	0.1	MPN/100 mL	10	1.0	0.1	MPN/100 mL
0	0	0	0	4	2	0	22
0	0	1	2	4	2	1	26
0	1	0	2	4	3	0	27
0	2	0	4	4	3	1	33
				4	4	0	34
1	0	0	2				
1	0	1	4	5	0	0	23
1	1	0	4	5	0	1	31
1	1	1	6	5	0	2	43
1	2	0	6	5	1	0	33
				5	1	1	46
2	0	0	5	5	1	2	63
2	0	1	7				
2	1	0	7	5	2	0	49
2	1	1	9	5	2	1	70
2	2	0	9	5	2	2	94
2	3	0	12	5	3	0	79
				5	3	1	110
3	0	0	8	5	3	2	140
3	0	1	11				
3	1	0	11	5	3	3	180
3	1	1	14	5	4	0	130
3	2	0	14	5	4	1	170
3	2	1	17	5	4	2	220
				5	4	3	280
4	0	0	13	5	4	4	350
4	0	1	17				
4	1	0	17	5	5	0	240
4	1	1	21	5	5	1	350
4	1	2	26	5	5	2	540
				5	5	3	920
				5	5	4	1600
				5	5	5	≥2400

Sample Volume, mL (10, 1.0, 0.1) | MPN/100 mL | Sample Volume, mL (10, 1.0, 0.1) | MPN/100 mL

(2) Include any positive results in dilutions higher than the selected series (0.001 mL dilution 1/5). This changes the selected series to 5-3-2.

(3) Using the first three columns on the reference chart, locate this series (5-3-2).

(4) Read the MPN value from the fourth column (140).

(5) On the chart the dilution series begins with 10 mL. For this test the series begins with 1.0 mL.

$$MPN / 100\,mL = 140\,MPN / 100\,mL \times \frac{10\,mL}{1\,mL}$$

$$= 1400\,MPN / 100mL$$

6.3.6.4 Membrane Filtration (MF) Technique

The *membrane filtration (MF) technique* is highly reproducible, can be used to test relatively large volumes of sample, and yields numerical results more rapidly than the multiple tube procedure. MTF is extremely useful in monitoring drinking water and a variety of natural waters. However, the MF technique has limitations, particularly when testing waters with high turbidity or non-coliform (background) bacteria. For such waters, or when the membrane filter technique has not been used previously, conducting parallel tests with the multiple tube fermentation technique to demonstrate applicability and comparability is desirable.

Note: As related to the membrane filter technique, the coliform group may be defined as comprising all aerobic and many facultative anaerobic, gram-negative, nonspore-forming, rod-shaped bacteria that develop a red colony with a metallic sheen within 24 hours at 35°C or an endo-type or nucleated colony without a metallic sheen. When verified these are classified as atypical coliform colonies. When purified cultures of coliform bacteria are tested, they produce a negative cytochrome oxidase (CO) and positive beta-galactosidase (ONPG) reaction. Generally, all red, pink, blue, white, or colorless colonies lacking sheen are considered non-coliforms by this technique.

The membrane filter technique uses a specially designed filter pad with uniformly sized pores (openings) small enough to prevent bacteria from entering the filter. A measured volume of sample is drawn through the filter pad by applying a partial vacuum. The special pad retains the bacteria on its surface, while allowing the water to pass through.

Note: In the membrane filter method, the number of coliforms is estimated by the number of colonies grown.

6.3.6.4.1 Membrane Filter (MF) Procedure

Note: When using the MF procedure for chlorinated effluents, before accepting it as an alternative, you must be able to demonstrate that it gives comparable information to that obtained by the multiple tube test.

The MF procedure uses an enriched lactose medium and incubation temperature of 44.5 +/– 0.2°C for selectivity, and gives 93% accuracy in differentiating between coliforms found in the feces of warm-blooded animals and those from other environmental sources. Because incubation temperature is critical, submerge waterproofed (plastic bag enclosures) MF cultures in a water bath for incubation at the elevated temperature, or use an appropriate, accurate solid heat sink incubator. Alternatively, use an equivalent incubator that will hold the 44.5°C temperature within 0.2°C (throughout the chamber) over a 24-hour period, while located in an environment of ambient air temperatures ranging from 5°C to 35°C.

Materials and Culture Medium:

(1) *M-FC medium:* The need for uniformity dictates the use of dehydrated media. Never prepare media from basic ingredients when suitable dehydrated media are available. Follow manufacturer's directions for rehydration. Commercially prepared media in liquid form (sterile ampule or other) also may be used, if known to give equivalent results.

(2) *Culture dishes:* Use tight-fitting plastic dishes because the MF cultures are submerged in a water bath during incubation. Enclose groups of fecal coliform cultures in plastic bags or seal individual dishes with waterproof (freezer) tape to prevent leakage during submersion.

(3) *Incubator:* The specificity of the fecal coliform test is related directly to the incubation temperature. Static air incubation may be a problem in some types of incubators, because of potential heat layering within the chamber and the slow recovery of temperature each time the incubator is opened during daily operations. To meet the need for greater temperature control use a water bath, a heat-sink incubator, or a properly designed and constructed incubator giving equivalent results. A temperature tolerance of 44.5 +/– 0.2°C can be obtained with most types of water baths equipped with a gable top for the reduction of heat and water losses. A circulating water bath is excellent, but may not be essential to this test, if the maximum permissible variation of 0.2°C in temperature can be maintained with other equipment.

Procedure:

(1) Sample Filtration

   a. Select a filter and aseptically separate it from the sterile package.

   b. Place the filter on the support plate with the grid side up.

   c. Place the funnel assembly on the support; secure as needed.

   d. Pour 100 mL of sample or serial dilution on to the filter, apply vacuum.

Note: The sample size and/or necessary serial dilution should produce a growth of 20–60 fecal coliform colonies on at least one filter. The selected dilutions must also be capable of showing permit excursions.

e. Allow all of the liquid to pass through the filter.

f. Rinse the funnel and filter with three portions (20–30 mL) of sterile, buffered dilution water. (Allow each portion to pass through the filter before the next addition.)

g. Remove the filter funnel and aseptically transfer the filter, grid side up, onto the prepared media.

Note: Filtration units should be sterile at the start of each filtration series and should be sterilized again if the series is interrupted for 30 minutes or more. A rapid interim sterilization can be accomplished by two minutes exposure to ultraviolet (UV) light, flowing steam, or boiling water.

(2) Incubation

a. Place absorbent pad into culture dish using sterile forceps.

b. Add 1.8 to 2.0 mL M-FC media to the absorbent pad.

c. Discard any media not absorbed by the pad.

d. Filter sample through sterile filter.

e. Remove filter from assembly and place on absorbent pad (grid up).

f. Cover culture dish.

g. Seal culture dishes in a weighted plastic bag.

h. Incubate filters in a water bath for 24 hours at 44.5 +/– 0.2°C.

6.3.6.4.2 Colony Counting

Upon completion of the incubation period, the surface of the filter will have growths of both fecal coliform and non-fecal coliform bacterial colonies. The fecal coliform will appear blue in color, while non-fecal coliform colonies will appear gray or cream colored.

When counting the colonies the entire surface of the filter should be scanned using a 10×–15× binocular, wide field dissecting microscope.

The desired range of colonies for the most valid fecal coliform determination is 20 to 60 colonies per filter. If multiple sample dilutions are used for the test, counts for each filter should be recorded on the laboratory data sheet.

a. Too many colonies

Filters that show a growth over the entire surface of filter with no individually identifiable colonies should be recorded as "confluent growth."

Filters that show a very high number of colonies (greater than 200) should be recorded as TNTC (too numerous to count).

   b. Not enough colonies

If no single filter meets the desired minimum colony count (20 colonies), the sum of the individual filter counts and the respective sample volumes can be used in the formula to calculate the colonies/100 mL.

Note: In each of these cases, adjustments in sample dilution volumes should be made to ensure future tests meet the criteria for obtaining a valid test result.

6.3.6.4.3 Calculation

The fecal coliform density can be calculated using the following formula:

$$\text{Colonies}/100\,\text{mL} = \frac{\text{Colonies Counted}}{\text{Sample Volume, mL}} \times 100\,\text{mL} \qquad (6.2)$$

Example 6.2

Using the data shown below, calculate the colonies per 100 mL for the influent and effluent samples noted.

Sample Location	Influent Sample Dilution			Effluent Sample Dilutions		
mL of Sample	1.0	0.1	0.01	10	1.0	0.1
Colonies Counted	97	48	16	10	5	3

Step 1: Influent Sample

Select the influent sample filter which has a colony count in the desired range (20 to 60). Since one filter meets this criteria, the remaining influent filters that did not meet the criteria are discarded.

$$\text{Colonies}/100\,\text{mL} = \frac{48\,\text{Colonies}}{0.1\,\text{mL}} \times 100$$

$$= 48,000\,\text{colonies}/100\,\text{mL}$$

Step 2: Effluent Sample

Since none of the filters for the effluent sample meets the minimum test requirement, the colonies/100 mL must be determined by totaling the colonies on each filter and the sample volumes used for each filter.

$$\text{Total Colonies} = 10 + 5 + 3$$

$$= 18 \text{ colonies}$$

$$\text{Total Sample} = 10.0 \text{ mL} + 1.0 \text{ mL} + 0.1 \text{ mL}$$

$$= 11.1 \text{ mL}$$

$$\text{Colonies}/100 \text{ mL} = \frac{18 \text{ colonies}}{11.1 \text{ mL}} \times 100$$

$$= 162 \text{ colonies}/100 \text{ mL}$$

Note: The USEPA criterion for fecal coliform bacteria in bathing waters is a logarithmic mean of 200 per 100 mL, based on a minimum of five samples taken over a 30-day period, with not more than 10% of the total samples exceeding 400 per 100 mL. Since shellfish may be eaten without being cooked, the strictest coliform criterion applies to shellfish cultivation and harvesting. The USEPA criterion states that the mean fecal coliform concentration should not exceed 14 per 100 mL, with not more than 10% of the samples exceeding 43 per 100 mL.

6.3.6.4.4 Interferences

Large amounts of turbidity, algae, or suspended solids may interfere with this technique by blocking the filtration of the sample through the membrane filter. Dilution of these samples to prevent this problem may make the test inappropriate for samples with low fecal coliform densities, since the sample volumes after dilution may be too small to give representative results. The presence of large amounts of non-coliform group bacteria in the samples may also prohibit the use of this method.

6.3.6.4.5 Geometric Mean Calculation

Many NPDES discharge permits require fecal coliform testing. Results for fecal coliform testing must be reported as a *geometric mean (average)* of all the test results obtained during a reporting period. A geometric mean, unlike an arithmetic mean or average, dampens the effect of very high or low values that otherwise might cause a nonrepresentative result.

Note: Current regulatory requirements prohibit the reporting of zero MPN or colonies. If the test result does not produce any positive results or colonies, the test result must be reported as <1 (less than 1). In cases where test results are reported as zero or <1, a value of "1" should be used in the calculation of the geometric mean. This substitution does not affect the result of the calculation; it just ensures that the data is entered into the calculation in a usable form.

Calculation of the geometric mean can be performed by either of two methods. Both methods require a calculator capable of performing more advanced calculations. The first method requires a calculator which is capable of determining the nth root of a number (n = the number of values used in the calculation). The general formula for this method of calculating the geometric mean is:

$$\text{Geometric Mean} = \sqrt[n]{X_1 \times X_2 \times \cdots \times X_n} \qquad (6.3)$$

This equation states that the geometric mean can be found by multiplying all of the data points for the given reporting period together and taking the nth root of this product.

Example 6.3

Given the data in the chart below, determine the geometric mean using the nth root method.

Solution:

$$\text{Geometric Mean} = \sqrt[4]{5 \times 7 \times 90 \times 1000}$$

$$= \sqrt[4]{3{,}150{,}000}$$

$$= 42 \text{ colonies} / 100 \text{ mL}$$

The second method for calculation of the geometric mean requires a calculator which can compute logarithms (log) and antilogarithms (antilog).

$$\text{Geometric Mean} = \text{antilog}\left(\frac{\log X_1 + \log X_2 + \log X_3 + \ldots + \log X_n}{N_1 \text{ Number of Tests}}\right) \qquad (6.4)$$

Procedure:

(1) If there are any reported values of 0, replace them with <1.

(2) Using the calculator, determine the logarithm of each test result.

(3) Add the logarithms of all of the test results.

(4) Divide the sum by the number of test results (N).

(5) Enter this number into the calculator.

(6) Press '2nd' or 'INV' then 'LOG.'

(7) The calculator displays the geometric mean.

Example 6.4

Given the following data, determine the geometric mean: week 1 = 12 MPN/100 mL, week 2 = 28 MPN/100 mL, week 3 = 37 MPN/100 mL, week 4 = 25 MPN/100 mL.

Solution:

$$\text{Geometric Mean} = \text{antilog}\left(\frac{1.079182 + 1.447158 + 1.568202 + 1.397940}{4}\right)$$

$$= \text{antilog}\left(\frac{5.492481}{4}\right)$$

$$= \text{antilog} \, 1.373120$$

$$= 23.6 \, \text{MPN} / 100 \, \text{mL}$$

6.4 VIRUSES

Viruses are parasitic particles—the smallest living infectious agents known. Since they are parasitic entities, they are not cellular (have no nucleus, cell membrane, or cell wall), are unable to multiply or adapt to the hostile water environment, and lack a living host. They multiply only within living cells (hosts) and are totally inert outside of living cells, but still can survive in the environment. Viruses differ from living cells in at least three ways: (1) they are unable to reproduce independently of cells and cannot carry out cell division; (2) they possess only one type of nucleic acid, either DNA or RNA; and (3) they have a simple acellular organization (Prescott et al., 1993).

Viruses can infect humans (it only takes a single virus particle to infect a host) through recently contaminated drinking water, the relativity of time being connected with the survival ability of the virus in natural and man-made hostile environments.

At present, over 100 virus types are known to occur in human feces, and an infected person may excrete as many as 10^6 infectious particles in 1 g of feces. Thus the potential for contamination is very great. Of those contacting viruses, only a small percentage are infected, and of those infected, only about 2% may become recognizably ill. Assuming a 1% infection rate and a 2% illness rate, this means that one in every 5000 persons coming in contact with a virus becomes ill, a very high rate if water is contaminated with fecal matter (Tchobanoglous and Schroeder, 1987).

Isolation of viruses has improved considerably during the past 40 years. They can be controlled by chlorination, but at much higher levels than are necessary to kill bacteria. The viruses examined by the drinking water practitioner are practically limited to enteric viruses (infections of the intestinal tract). Some viruses that may be transmitted by water include infectious hepatitis, adenovirus, polio, coxsackie, echo, and Norwalk agent. A virus that infects a bacterium is called a *bacteriophage*.

6.4.1 BACTERIOPHAGE

Lewis Thomas (1974) points out that when humans "catch diphtheria it is a virus infection, but not of us." In other words, when humans are infected by the virus causing diphtheria, the bacterium is really infected—humans simply "blundered into someone else's accident" (p. 76). The toxin of diphtheria bacilli is produced by the organisms that have been infected with a bacteriophage.

A bacteriophage (phage) is any viral organism whose host is a bacterium. Most bacteriophage research has been carried out on the bacterium *Escherichia coli*, which is one of the gram-negative bacteria that water specialists are concerned with because it is a typical coliform.

6.4.2 INDICATOR VIRUSES

Considerable research has been accomplished in the last thirty years in the attempt to determine certain viruses as indicator viruses. In their book *Drinking Water and Health*, the National Academy of Sciences (1977) reached the following conclusions:

(1) The presence of infecting virus in drinking water is a potential hazard to public health, and no valid basis exists upon which a no-effect concentration of viral contamination in finished drinking water might be established.

   Note: This statement in no way should be interpreted to mean that viruses are not removed by conventional treatment and disinfection. On the contrary, drinking water produced by an effective conventional treatment and distributed after disinfection is expected to have significantly reduced concentrations of viruses inactivated by the treatment.

(2) Continued testing for viral contamination of potable water should be

carried out with the facilities and skills of a wide variety of research establishments, both inside and outside of government, and methodology for virus testing should be improved.

(3) The bacteriological monitoring methods currently prescribed or recommended in this report (coliform count and standard plate count) are the best indicators available today for routine use in evaluating the presence in water of intestinal pathogens, including viruses.

In 1987, the USEPA concluded that measuring the level of enteric viruses in drinking water is not economically or technologically feasible for the following reasons: Presently acceptable methods require levels of expertise that utility personnel normally do not possess, and/or the methods would be too expensive if analyzed by private laboratories. Validation procedures have not yet been established. Continuous monitoring would be required. Monitoring also does not provide advance notice to assure the safety of drinking water at the consumer's tap (De Zuane, 1997).

6.5 PROTOZOA

The *protozoa* (first animals) are a large group (more than 50,000 known species) of eucaryotic organisms that have adapted a form or cell to serve as the entire body. In fact, all protozoans are single-celled organisms. Typically, they lack cell walls, but have a plasma membrane that is used to take in food and discharge waste. They can exist as solitary or independent organisms (for example, the stalked ciliates such as *Vorticella* sp.) or they can colonize (like the sedentary *Carchesium* sp.). Protozoa are microscopic and get their name because they employ the same type of feeding strategy as animals. Most are harmless, but some are parasitic. Some forms have two life stages: active trophozoites (capable of feeding) and dormant cysts.

As unicellular eucaryotes, protozoa can't be easily defined because they are diverse, and in most cases, only distantly related to each other (Patterson and Hedley, 1992). Each protozoan is a complete organism and contains the facilities for performing the body functions for which vertebrates have many organ systems.

6.5.1 PATHOGENIC PROTOZOA

Certain types of protozoans can cause disease. Of particular interest to the drinking water practitioner are the *Entamoeba histolytica* (amebic dysentery and amebic hepatitis), *Giardia lamblia* (Giardiasis), *Cryptosporidium* (cryptosporidiosis), and the emerging *Cyclospora* (cyclosporosis). Sewage contamination transports eggs, cysts, and oocysts of parasitic protozoa and helminths (tapeworms, hookworms, etc.) into raw supplies, leaving water treatment and disinfection as the means by which to diminish the danger of contaminated water for the consumer.

To prevent the occurrence of *Giardia* and *Cryptosporidium* spp. in surface water supplies, and to address increasing problems with waterborne diseases, the USEPA implemented its Surface Water Treatment Rule (SWTR) on 29 June 1989. The rule requires both filtration and disinfection of all surface water supplies as a means of primarily controlling *Giardia* spp. and enteric viruses. Since implementation of its Surface Water Treatment Rule, the USEPA has also recognized that *Cryptosporidium* spp. is an agent of waterborne disease (Badenock, 1990). In 1996, in its next series of surface water regulations, the USEPA included *Cryptosporidium*.

To test the need for and the effectiveness of the USEPA's Surface Water Treatment Rule, LeChevallier et al. (1991) conducted a study on the occurrence and distribution of *Giardia* and *Cryptosporidium* organisms in raw water supplies to 66 surface water filter plants. These plants were located in fourteen states and one Canadian province. A combined immunofluorescence test indicated that cysts and oocysts were widely dispersed in the aquatic environment. *Giardia* spp. were detected in more than 80% of the samples. *Cryptosporidium* spp. were found in 85% of the sample locations. Taking into account several variables, *Giardia* or *Cryptosporidium* spp. were detected in 97% of the raw water samples. After evaluating their data, the researchers came to the conclusion that the Surface Water Treatment Rule may have to be upgraded (subsequently, it has been) to require additional treatment.

6.5.1.1 *Giardia*

Giardia (gee-ar-dee-ah) *lamblia* (also known as hiker's/traveler's scourge or disease) is a microscopic parasite that can infect warm-blooded animals and humans. Although *Giardia* was discovered in the 19th century, not until 1981 did the World Health Organization (WHO) classify *Giardia* as a pathogen. *Giardia* is protected by an outer shell called a cyst that allows it to survive outside the body for long periods of time. If viable cysts are ingested, *Giardia* can cause the illness known as *giardiasis,* an intestinal illness which can cause nausea, anorexia, fever, and severe diarrhea. The symptoms last only for several days, and the body can naturally rid itself of the parasite in one to two months. However, for individuals with weakened immune systems, the body often cannot rid itself of the parasite without medical treatment.

In the United States, *Giardia* is the most commonly identified pathogen in waterborne disease outbreaks. Contamination of a water supply by *Giardia* can occur in two ways: (1) by the activity of animals in the watershed area of the water supply; or (2) by the introduction of sewage into the water supply. Wild and domestic animals are major contributors in contaminating water supplies. Studies have also shown that, unlike many other pathogens, *Giardia* is not host-specific. In short, *Giardia* cysts excreted by animals can infect and cause illness in humans. Additionally, in several major outbreaks of waterborne diseases, the *Giardia* cyst source was sewage contaminated water supplies.

Waterborne *Giardia*, however, can be effectively controlled by treating the water supply. Chlorine and ozone are examples of two disinfectants known to effectively kill *Giardia* cysts. Filtration of the water can also effectively trap and remove the parasite from the water supply. The combination of disinfection and filtration is the most effective water treatment process available today for prevention of *Giardia* contamination.

In drinking water, *Giardia* is regulated under the Surface Water Treatment Rule. Although the SWTR does not establish a Maximum Contaminant Level (MCL) for *Giardia*, it does specify treatment requirements to achieve at least 99.9% removal and/or inactivation of *Giardia*. This regulation requires that all drinking water systems using surface water or groundwater under the influence of surface water must disinfect and filter the water. The Enhanced Surface Water Treatment Rule (ESWTR), which includes *Cryptosporidium* and further regulates *Giardia*, was established in December 1996.

6.5.1.1.1 Giardiasis[6]

During the past fifteen years, *giardiasis* has been recognized as one of the most frequently occurring waterborne diseases in the United States. *Giardia lamblia* cysts have been discovered in the U.S. in places as far apart as Estes Park, Colorado (near the Continental Divide); Missoula, Montana; Wilkes-Barre, Scranton, and Hazleton, Pennsylvania; and Pittsfield and Lawrence, Massachusetts, just to name a few.

Giardiasis is characterized by intestinal symptoms that usually last one week or more and may be accompanied by one or more of the following: diarrhea, abdominal cramps, bloating, flatulence, fatigue, and weight loss. Although vomiting and fever are commonly listed as relatively frequent symptoms, they have not been commonly reported by people involved in waterborne outbreaks in the U.S.

While most *Giardia* infections persist only for one or two months, some people undergo a more chronic phase, which can follow the acute phase or may become manifest without an antecedent acute illness. The chronic phase is characterized by loose stools and increased abdominal gassiness with

---

[6]Much of the information contained in this section is from the U.S. Centers for Disease Control, *Giardiasis*, by D. D. Juranek, 1995.

cramping, flatulence, and burping. Fever is not common, but malaise, fatigue, and depression many ensue (Weller, 1985). For a small number of people, the persistence of infection is associated with the development of marked malabsorption and weight loss (Weller, 1985). Similarly, lactose (milk) intolerance can be a problem for some people. This can develop coincidentally with the infection or be aggravated by it, causing an increase in intestinal symptoms after ingestion of milk products.

Some people may have several of these symptoms without evidence of diarrhea or have only sporadic episodes of diarrhea every three or four days. Still others may not have any symptoms at all. Therefore, the problem may not be whether you are infected with the parasite or not, but how harmoniously you both can live together or how to get rid of the parasite (either spontaneously or by treatment) when the harmony does not exist or is lost.

Three prescription drugs are available in the United States to treat giardiasis: quinacrine, metronidazole, and furazolidone. In a recent review of drug trials in which the efficacies of these drugs were compared, quinacrine produced a cure in 93% of patients, metronidazole cured 92%, and furazolidone cured about 84% of patients (Davidson, 1984). Quinacrine is generally the least expensive of the anti-*Giardia* medications, but it often causes vomiting in children younger than five years old. Although the treatment of giardiasis is not an FDA-approved indication for metronidazole, the drug is commonly used for this purpose. Furazolidone is the least effective of the three drugs, but is the only anti-*Giardia* medication that comes in liquid preparation, which makes it easier to deliver the exact dose to small children, and the most convenient dosage form for children who have difficulty taking pills.

Giardiasis occurs worldwide. In the U.S., *Giardia* is the parasite most commonly identified in stool specimens submitted to state laboratories for parasitologic examination. During a three-year period, approximately 4% of one million stool specimens submitted to state laboratories tested positive for *Giardia* (CDC, 1979). Other surveys have demonstrated *Giardia* prevalence rates ranging from 1 to 20%, depending on the location and ages of persons studied. Giardiasis ranks among the top 20 infectious diseases that cause the greatest morbidity in Africa, Asia, and Latin America (Walsh and Warren, 1979); it has been estimated that about two million infections occur per year in these regions (Walsh, 1981).

People who are at highest risk for acquiring *Giardia* infection in the U.S. may be placed into five major categories:

(1) People in cities whose drinking water originates from streams or rivers, and whose water treatment process does not include filtration, or where filtration is ineffective because of malfunctioning equipment
(2) Hikers/campers/outdoors people

(3) International travelers

(4) Children who attend day-care centers, day-care center staff, and parents and siblings of children infected in day-care centers

(5) Homosexual men

People in categories 1, 2, and 3 have in common the same general source of infections, i.e., they acquire *Giardia* from fecally contaminated drinking water. The city resident usually becomes infected because the municipal water treatment process does not include the filter necessary to physically remove the parasite from the water. The number of people in the U.S. at risk (i.e., the number who receive municipal drinking water from unfiltered surface water) is estimated to be 20 million. International travelers may also acquire the parasite from improperly treated municipal waters in cities or villages in other parts of the world, particularly in developing countries. In Eurasia, only travelers to Leningrad appear to be at increased risk. In prospective studies, 88% of U.S. and 35% of Finnish travelers to Leningrad who had negative stool tests for *Giardia* on departure to the Soviet Union developed symptoms of giardiasis and had positive tests for *Giardia* after they returned home (Brodsky et al., 1974; Jokipii et al., 1974). With the exception of visitors to Leningrad, however, *Giardia* has not been implicated as a major cause of traveler's diarrhea—it has been detected in fewer than 2% of travelers who develop diarrhea. However, hikers and campers risk infection every time they drink untreated raw water from a stream or river.

Persons in categories 4 and 5 become exposed through more direct contact with feces of an infected person: exposure to solid diapers of an infected child (day-care center-associated cases), or through direct or indirect anal-oral sexual practices in the case of homosexual men.

Although community waterborne outbreaks of giardiasis have received the greatest publicity in the U.S. during the past decade, about half of the *Giardia* cases discussed with staff of the Centers for Disease Control over a three-year period had a day-care center exposure as the most likely source of infection. Numerous outbreaks of *Giardia* in day-care centers have been reported in recent years. Infection rates for children in day-care center outbreaks range from 21 to 44% in the U.S. and from 8 to 27% in Canada (Black et al., 1981; Pickering et al., 1984; Pickering et al., 1981; Sealy et al., 1983; Keystone et al., 1984; Keystone et al., 1978). The highest infection rates are usually observed in children who wear diapers (one to three years of age). In a study of 18 randomly selected day-care centers in Atlanta (CDC, unpublished data), 10% of diapered children were found infected. Transmission from this age group to older children, day-care staff, and household contacts is also common. About 20% of parents caring for an infected child become infected.

Local health officials and managers of water utility companies need to realize that sources of *Giardia* infection other than municipal drinking water exist. Armed with this knowledge, they are less likely to make a quick (and sometimes wrong) assumption that a cluster of recently diagnosed cases in a city is related to municipal drinking water. Of course, drinking water must not be ruled out as a source of infection when a larger than expected number of cases is recognized in a community, but the possibility that the cases are associated with a day-care center outbreak, drinking untreated stream water, or international travel should also be entertained.

To understand the finer aspects of *Giardia* transmission and strategies for control, the drinking water practitioner must become familiar with several aspects of the parasite's biology. Two forms of the parasite exist: a *trophozoite* and a *cyst*, both of which are much larger than bacteria (see Figure 6.4). Trophozoites live in the upper small intestine where they attach to the

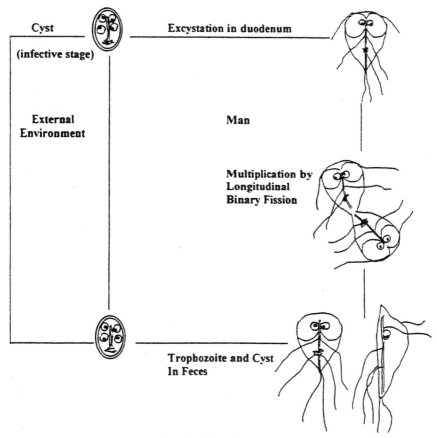

Cyst

(infective stage)

Excystation in duodenum

External Environment

Man

Multiplication by Longitudinal Binary Fission

Trophozoite and Cyst In Feces

Figure 6.4 Life cycle of *Giardia lamblia. Source:* CDC.

intestinal wall by means of a disc-shaped suction pad on their ventral surface. Trophozoites actively feed and reproduce at this location. At some time during the trophozoite's life, it releases its hold on the bowel wall and floats in the fecal stream through the intestine. As it makes this journey, it undergoes a morphologic transformation into an egglike structure called a cyst. The cyst (about 6 to 9 micrometers in diameter × 8 to 12 micrometers—1/100 millimeter—in length) has a thick exterior wall that protects the parasite against the harsh elements that it will encounter outside the body. This cyst form of the parasite is infectious to other people or animals. Most people become infected either directly (by hand-to-mouth transfer of cysts from the feces of an infected individual) or indirectly (by drinking feces-contaminated water). Less common modes of transmission included ingestion of fecally contaminated food and hand-to-mouth transfer of cysts after touching a fecally contaminated surface. After the cyst is swallowed, the trophozoite is liberated through the action of stomach acid and digestive enzymes and becomes established in the small intestine.

Although infection after the ingestion of only one *Giardia* cyst is theoretically possible, the minimum number of cysts shown to infect a human under experimental conditions is ten (Rendtorff, 1954). Trophozoites divide by binary fission about every 12 hours. What this means in practical terms is that if a person swallowed only a single cyst, reproduction at this rate would result in more than one million parasites 10 days later, and one billion parasites by day 15.

The exact mechanism by which *Giardia* causes illness is not yet well understood, but is not necessarily related to the number of organisms present. Nearly all of the symptoms, however, are related to dysfunction of the gastrointestinal tract. The parasite rarely invades other parts of the body, such as the gall bladder or pancreatic ducts. Intestinal infection does not result in permanent damage.

Data reported by the CDC indicate that *Giardia* is the most frequently identified cause of diarrheal outbreaks associated with drinking water in the United States. The remainder of this section is devoted to waterborne transmissions of *Giardia*. Waterborne epidemics of giardiasis are a relatively frequent occurrence. In 1983, for example, *Giardia* was identified as the cause of diarrhea in 68% of waterborne outbreaks in which the causal agent was identified (CDC, 1984). From 1965 to 1982, more than 50 waterborne outbreaks were reported (Craun, 1984). In 1984, about 250,000 people in Pennsylvania were advised to boil drinking water for six months because of *Giardia*-contaminated water.

Many of the municipal waterborne outbreaks of *Giardia* have been subjected to intense study to determine their cause. Several general conclusions can be made from data obtained in those studies. Waterborne transmission of *Giardia* in the U.S. usually occurs in mountainous regions where community drinking water obtained from clear running streams is chlorinated but not filtered before distribution. Although mountain streams appear to be clean, fecal contamination upstream by human residents or visitors, as well as by *Giardia*-infected animals such as beavers, has been well documented. Water obtained from deep wells is an unlikely source of *Giardia* because of the natural filtration of water as it percolates through the soil to reach underground cisterns. Well-water sources that pose the greatest risk of fecal contamination are poorly constructed or improperly located ones. A few outbreaks have occurred in towns that included filtration in the water treatment process, where the filtration was not effective in removing *Giardia* cysts because of defects in filter construction, poor maintenance of the filter media, or inadequate pretreatment of the water before filtration. Occasional outbreaks have also occurred because of accidental cross-connections between water and sewage systems.

From these data we conclude that two major ingredients are necessary for waterborne outbreak. *Giardia* cysts must be present in untreated source water, and the water purification process must either fail to kill or fail to remove *Giardia* cysts from the water.

Though beavers are often blamed for contaminating water with *Giardia* cysts, that they are responsible for introducing the parasite into new areas seems unlikely. Far more likely is that they are also victims: *Giardia* cysts may be carried in untreated human sewage discharged into the water by small-town sewage disposal plants or originate from cabin toilets that drain directly into streams and rivers. Backpackers, campers, and sports enthusiasts may also deposit *Giardia*-contaminated feces in the environment, which are subsequently washed into streams by rain. In support of this concept is a growing amount of data that indicate a higher *Giardia* infection rate in beavers living downstream from U.S. National Forest campgrounds, compared with a near zero rate of infection in beavers living in more remote areas.

Although beavers may be unwitting victims of the *Giardia* story, they still play an important part in the transmission scheme, because they can (and probably do) serve as amplifying hosts. An *amplifying host* is one that is easy to infect, serves as a good habitat for the parasite to reproduce, and in the case of *Giardia,* returns millions of cysts to the water for every one ingested. Beavers are especially important in this regard, because they tend to defecate in or very near the water, which ensures that most of the *Giardia* cysts excreted are returned to the water.

The contribution of other animals to waterborne outbreaks of *Giardia* is less clear. Muskrats (another semiaquatic animal) have been found in several parts of the United States to have high infection rates (30 to 40%) (Frost et al., 1984). Recent studies have shown that muskrats can be infected with *Giardia* cysts obtained from humans and beavers. Occasional *Giardia* infections have been reported in coyotes, deer, elk, cattle, dogs, and cats (but not in horses and sheep) encountered in mountainous regions of the United States. Naturally-occurring *Giardia* infections have not been found in most other wild animals (bear, nutria, rabbit, squirrel, badger, marmot, skunk, ferret, porcupine, mink, raccoon, river otter, bobcat, lynx, moose, bighorn sheep) (Frost et al., 1984).

Scientific knowledge about what is required to kill or remove *Giardia* cysts from a contaminated water supply has increased considerably. For example, we know that cysts can survive in cold water (4°C) for at least two months, and that they are killed instantaneously by boiling water (100°C) (Frost et al., no date; Bingham et al., 1979). We do not know how long the cysts will remain viable at other water temperatures (e.g., at 0°C or in a canteen at 15–20°C), nor do we know how long the parasite will survive on various environment surfaces, e.g., under a pine tree, in the sun, on a diaper-changing table, or in carpets in a day-care center.

The effect of chemical disinfection (chlorine, for example) on the viability of *Giardia* cysts is an even more complex issue. The number of waterborne outbreaks of *Giardia* that have occurred in communities where chlorine was employed as a disinfectant demonstrates that the amount of chlorine used routinely for municipal water treatment is not effective against *Giardia* cysts. These observations have been confirmed in the laboratory under experimental conditions (Jarroll et al., 1979; Jarroll et al., 1980; Jarroll et al., 1981). This does not mean that chlorine does not work at all. It does work under certain favorable conditions. Without getting too technical, gaining some appreciation of the problem can be achieved by understanding a few of the variables that influence the efficacy of chlorine as a disinfectant.

(1) Water pH: at pH values above 7.5, the disinfectant capability of chlorine is greatly reduced.

(2) Water temperature: the warmer the water, the higher the efficacy. Chlorine does not work in ice-cold water from mountain streams.

(3) Organic content of the water: mud, decayed vegetation, or other suspended organic debris in water chemically combines with chlorine, making it unavailable as a disinfectant.

(4) Chlorine contact time: the longer *Giardia* cysts are exposed to chlorine, the more likely the chemical will kill them.

(5) Chlorine concentration: the higher the chlorine concentration, the more likely chlorine will kill *Giardia* cysts. Most water treatment facilities try to add enough chlorine to give a free (unbound) chlorine residual at the customer tap of 0.5 mg per liter of water.

The five variables above are so closely interrelated that an unfavorable occurrence in one can often be compensated for by improving another. For example, if chlorine efficacy is expected to be low because water is obtained from an icy stream, the chlorine contact time, chlorine concentration, or both could be increased. In the case of *Giardia*-contaminated water, producing safe drinking water with a chlorine concentration of 1 mg per liter and a contact time as short as 10 minutes might be possible—if all the other variables were optimal (i.e., pH of 7.0, water temperature of 25°C, and a total organic content of the water close to zero). On the other hand, if all of these variables were unfavorable (i.e., pH of 7.9, water temperature of 5°C, and high organic content), chlorine concentrations in excess of 8 mg per liter with several hours of contact time may not be consistently effective. Because water conditions and water treatment plant operations (especially those related to water retention time, and therefore, to chlorine contact time) vary considerably in different parts of the United States, neither the USEPA nor the CDC has been able to identify a chlorine concentration that would be safe yet effective against *Giardia* cysts under all water conditions. Therefore, the use of chlorine as a preventive measure against waterborne giardiasis generally has been used under outbreak conditions when the amount of chlorine and contact time have been tailored to fit specific water conditions and the existing operational design of the water utility.

In an outbreak, for example, the local health department and water utility may issue an advisory to boil water, may increase the chlorine residual at the consumer's tap from 0.5 mg/L to 1 or 2 mg/L, and if the physical layout and operation of the water treatment facility permit, increase the chlorine contact time. These are emergency procedures intended to reduce the risk of transmission until a filtration device can be installed or repaired or until an alternative source of safe water (a well, for example) can be made operational.

The long-term solution to the problem of municipal waterborne outbreaks of giardiasis involves improvements in and more widespread use of filters in the municipal water treatment process (see Chapter 11). The sand filters most commonly used in municipal water treatment today cost millions of dollars to install, which makes them unattractive for many small communities. The pore sizes in these filters are not sufficiently small to remove a *Giardia* (6 to 9 micrometers × 8 to 12 micrometers). For the sand filter to remove *Giardia* cysts from the water effectively, the water must receive some additional treatment before it reaches the filter. The flow of water through the filter bed must also be carefully regulated.

An ideal prefilter treatment for muddy water would include sedimentation (a holding pond where large suspended particles are allowed to settle out by the action of gravity; see Chapter 11) followed by flocculation or coagulation (the addition of chemicals such as alum or ammonium to cause microscopic particles to clump together). The large particles resulting from the flocculation/coagulation process, including *Giardia* cysts bound to other micropar-ticulates, are easily removed by the sand filter. Chlorine is then added to kill the bacteria and viruses that may escape the filtration process. If the water comes from a relatively clear source, chlorine may be added to the water before it reaches the filter.

As should be clear to you, successful operation of a complete water treatment works is a complex process that requires considerable training. Troubleshooting breakdowns or recognizing the potential problems in the system before they occur often requires the skills of an engineer. Unfortu-nately, most small water utilities with water treatment facilities that include filtration cannot afford the services of a full-time engineer. Filter operation or maintenance problems in such systems may not be detected until a *Giardia* outbreak is recognized in the community. The bottom line is that, although water filtration is the best that water treatment technology has to offer for municipal systems against waterborne giardiasis, it is not infallible. For municipal water filtration facilities to work properly, they must be properly constructed, operated, and maintained.

Whenever possible, persons in the out-of-doors should carry drinking water of known purity with them. When this is not practical, when water from streams, lakes, ponds, and other outdoor sources must be used, time should be taken to properly disinfect the water before drinking it.

Boiling water is one of the simplest and most effective ways to purify it. Boiling for one minute is adequate to kill *Giardia* as well as most other bacterial or viral pathogens likely to be acquired from drinking polluted water.

Disinfection of water with chlorine or iodine is considered less reliable than boiling for killing *Giardia*. However, we recognize that boiling drink-ing water is not practical under many circumstances. When boiling drinking water is not possible, chemical disinfectants such as iodine or chlorine should be used. This provides some protection against *Giardia,* and destroys most bacteria and viruses that cause illness. Iodine or chlorine concentrations of 8 mg/L (8 ppm) with a minimum contact time of 30 minutes are recommended. If the water is cold (less than 10°C or 50°F) a minimum contact time of 60 minutes is recommended. If a choice of disinfectants is available, use iodine. Iodine's disinfectant activity is less likely than chlo-rine's to be reduced by unfavorable water conditions, such as dissolved organic material in water or by water with a high pH.

Table 6.4 gives instructions for disinfecting water using household tinc-ture of iodine; Table 6.5 gives instructions for using chlorine bleach. If water is visibly dirty, it should first be strained through a clean cloth into a container to remove any sediment or floating matter. Then the water should be treated with chemicals as shown in Table 6.4 and Table 6.5.

TABLE 6.4. Iodine.

Tincture of iodine from the medicine chest or first aid kit can be used to treat water. Mix thoroughly by stirring or shaking water in container and let stand for 30 minutes.		
	Drops (0.05 mL) to be added per quart/liter	
Tincture of iodine (let stand for several hours or overnight)	Clear Water	Cold or Cloudy Water
2%	5	10

A variety of portable filter devices for field or individual use as well as some household filters are available for use against waterborne giardiasis. Manufacturer's data accompanying these filters indicate that some can remove particles the size of a *Giardia* cyst or smaller and may be capable of providing a source of safe drinking water for an individual or family during a waterborne outbreak. If carefully selected, such devices might also be useful in preventing giardiasis for international travelers, backpackers, campers, sportsmen, or persons who live or work in areas where water is known to be contaminated.

Unfortunately, very few published reports in the scientific literature detail both the methods used and the results of tests employed to evaluate the efficacy of these filters against *Giardia*. Until more published experimental data become available, consumers should look for a few commonsense indications when selecting a portable or household filter.

The first thing to consider is the filter media. Filters relying solely on ordinary or silver-impregnated carbon or charcoal should be avoided, be-

TABLE 6.5. Chlorine.

Liquid chlorine bleach used for laundry usually has 4% to 6% available chlorine. Read the label to find the percentage of chlorine in the solution and follow the treatment schedule below.		
Mix thoroughly by stirring or shaking water in container for 30 minutes. Let stand for several hours or overnight. A slight chlorine odor should be detectable in the water.		
Available chlorine	Clear Water	Cold or Cloudy Water
1%	10	20
4% to 6%	2	4
7%to 10%	1	2
Unknown	10	20

cause they are not intended to prevent, destroy, or repel microorganisms. Their principal use is to remove undesirable chemicals, odors, and very large particles such as rust or dirt.

Some filters rely on chemicals such as iodide-impregnated resins to kill *Giardia*. While properly designed and manufactured iodide-impregnated resin filters have been shown to kill many species of bacteria and virus present in human feces, their efficacy against *Giardia* cysts is less well-established. The principle under which these filters operate is similar to that achieved by adding the chemical disinfectant iodine to water, except that the microorganisms in the water pass over the iodide-impregnated disinfectant as the water flows through the filter.

While the disinfectant activity of iodide is not as readily affected as chlorine by water pH or organic content, iodide disinfectant activity is markedly reduced by cold water temperatures. Experiments on *Giardia* indicate that many of the cysts in cold water (4°C) remain viable after passage through filters containing tri-iodide or penta-iodide disinfectant (Marchin et al., 1983). Longer contact time (compared to that required to kill bacteria) is required when using chemical filters to process cold water for *Giardia* protection. Presently available chemical filters are also not recommended for muddy or very turbid water. Note that filters relying solely on chemical action usually give no indication to the user when disinfectant activity has been depleted.

The so-called microstrainer types of filters are true filters. Manufacturer data accompanying these filters indicate that some have a sufficiently small pore size to physically restrict the passage of some microorganisms through the filter. The types of filter media employed in microstraining filters include orlon, ceramic, and proprietary materials. Theoretically, a filter having an absolute pore size of less than 6 micrometers might be able to prevent *Giardia* cysts of 8 to 10 micrometers in diameter from passing. When used as a water sampling device during community outbreaks, portable filters in the 1 to 3 micrometer range more effectively removed *Giardia* cysts from raw water than filters with larger pore sizes. For effective removal of bacterial or viral organisms that cause disease in humans, microstraining filters with pore sizes of less than 1 micrometer are advisable. However, the smaller the pores, the more quickly the filters will tend to clog. To obtain maximum filter life, and as a matter of reasonable precaution, the cleanest available water source should always be used. Keep in mind, however, that even sparkling, clear mountain streams can be heavily contaminated.

The second condition to examine when choosing a filter includes considering whether the filter element can be cleaned or replaced without posing a significant health hazard to the user, because infectious organisms can be concentrated on the filter element/media itself. Properly engineered portable filters should minimize the possibility of contaminating the "clean water side" of the filter with contaminated water during replacement or cleaning of the filter element. Since filters used in the field are often rinsed or "cleaned" in a stream or river that may be contaminated, this is especially important for recreational outdoor use.

6.5.1.2 *Cryptosporidium*

In 1907, when Ernest E. Tyzzer recognized, described, and published an account of a parasite he frequently found in the gastric glands of laboratory mice, he and his new discovery were hidden in virtual anonymity—just another scientist quietly going about his normal, tedious, out-of-the-lime-light research, buried in obscurity. Initially, his studies focused on describing the asexual and sexual stages and spores (oocysts), each with a specialized attachment organelle, and noted that spores were excreted in the feces (Tyzzer, 1907).

Tyzzer identified the parasite as a sporozoan, but of uncertain taxonomic status; he named it *Cryptosporidium muris*. Later, in 1910, after more detailed study, he proposed *Cryptosporidium* as a new genus and *C. muris* as the type species. Amazingly, except for developmental stages, Tyzzer's original description of the life cycle (see Figure 6.5) was later confirmed by electron microscopy. Later, in 1912, Tyzzer described a new species, *Cryptosporidium parvum*.

For almost 50 years, Tyzzer's discovery of the genus *Cryptosporidium* (because it appeared to be of no medical or economic importance) remained (like himself) relatively obscure. However, slight rumblings of the genus' importance were felt in the medical community when Slavin (1955) wrote about a new species, *Cryptosporidium meleagridis*, associated with illness and death in turkeys. Interest remained slight even when *Cryptosporidium* was found to be associated with bovine diarrhea (Panciera et al., 1971).

Not until 1982 did worldwide interest focus in on the study of organisms in the genus *Cryptosporidium*. During this time frame, the medical community and other interested parties were beginning to attempt a full-scale, frantic effort to find out as much as possible about Acquired Immune Deficiency Syndrome (AIDS). The CDC reported that 21 AIDS-infected males from six large cities in the U.S. had severe protracted diarrhea caused by *Cryptosporidium*.

However, 1993 was when the "bug—the pernicious parasite *Cryptosporidium*—made Milwaukee famous" (Mayo Foundation, 1996).

Note: The *Cryptosporidium* outbreak in Milwaukee caused the deaths of 100 people—the largest episode of waterborne disease in the U.S. in the 70 years since health officials began tracking such outbreaks.

Today we know that the massive waterborne outbreak in Milwaukee, Wisconsin (more than 400,000 persons developed acute and often prolonged

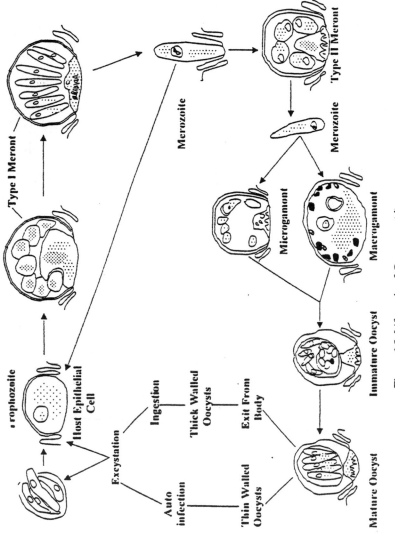

Figure 6.5 Life cycle of *Cryptosporidium parvum*.

Type I Meront

Merozoite

Type II Meront

Merozoite

Microgamont

Macrogamont

Immature Oocyst

Mature Oocyst

Thin Walled Oocysts

Thick Walled Oocysts

Auto infection

Ingestion

Exit From Body

Excystation

Host Epithelial Cell

rophozoite

diarrhea or other gastrointestinal symptoms), increased interest in *Cryptosporidium* at an exponential level. The Milwaukee Incident spurred both public interest and the interest of public health agencies, agricultural agencies and groups, environmental agencies and groups, and suppliers of drinking water. This increase in interest level and concern has spurred on new studies of *Cryptosporidium* with emphasis on developing methods for recovery, detection, prevention, and treatment (Fayer et al., 1997).

The USEPA has become particularly interested in this "new" pathogen. For example, in the reexamination of regulations on water treatment and disinfection, the USEPA issued MCLG and CCL for *Cryptosporidium*. The similarity to *Giardia lamblia* and the necessity to provide an efficient conventional water treatment capable of eliminating viruses at the same time forced the USEPA to regulate the surface water supplies in particular. The proposed "Enhanced Surface Water Treatment Rule" (ESWTR) included regulations from watershed protection to specialized operation of treatment plants (certification of operators and state overview) and effective chlorination. Protection against *Cryptosporidium* included control of waterborne pathogens such as *Giardia* and viruses (De Zuane, 1997).

6.5.1.2.1 The Basics of Cryptosporidium

Recently, a colleague asked us to give her the basic information on *Cryptosporidium*. We provided her with the following information.[7]

Cryptosporidium (crip-toe-spor-ID-ee-um) is one of several single-celled protozoan genera in the phylum Apircomplexa (all referred to as coccidia). *Cryptosporidium* along with other genera [*Eimeria, Isospora*, and *Cyclospora* (see Section 6.5.1.3)] in the phylum Apircomplexa develop in the gastrointestinal tract of vertebrates through all of their life cycle—in short, they live in the intestines of animals and people. This microscopic pathogen causes a disease called *cryptosporidiosis* (crip-toe-spor-id-ee-O-sis).

The dormant (inactive) form of *Cryptosporidium* called an oocyst (O-o-sist) is excreted in the feces (stool) of infected humans and animals. The tough-walled oocysts survive under a wide range of environmental conditions.

Several species of *Cryptosporidium* were incorrectly named after the host in which they were found; subsequent studies have invalidated many species. At the present time, eight valid species of *Cryptosporidium* (see Table 6.6) have been named.

Upton (1997) reports that *C. muris* infects the gastric glands of laboratory rodents and several other mammalian species, but (even though several texts

---

[7]Much of the information contained in this section is from USDA, Cryposporidium: *A Waterborne Pathogen*, provided by USDA Water Quality Cooperation Extension, Cornell U., 1998.

TABLE 6.6. Valid Named Species of *Cryptosporidium.*

Species	Host
C. baileyi	chicken
C. felis	domestic cat
C. meleagridis	turkey
C. murishouse	house mouse
C. nasorum	fish
C. parvum	house mouse
C. serpentis	corn snake
	rat snake
	Madagascar boa
C. wrairi	guinea pig

Source: Adapted from Fayer et al., "The General Biology of *Cryptosporidium.*" In *Cryptosporidium and Cryptosporidiosis,* ed. R. Fayer. Boca Raton, Florida: CRC Press, 1997.

state otherwise) is not known to infect humans. However, *C. parvum* infects the small intestine of an unusually wide range of mammals, including humans, and is the zoonotic species responsible for human cryptosporidiosis.

Upton goes on to explain that in most mammals *C. parvum* is predominately a parasite of neonate (newborn) animals. He points out that even though exceptions occur, older animals generally develop poor infections, even when unexposed previously to the parasite. Humans are the one host that can be seriously infected at any time in their lives, and only previous exposure to the parasite results in either full or partial immunity to challenge infections.

Oocysts are present in most surface bodies of water (e.g., lakes and rivers) across the U.S., many of which supply public drinking water. Oocysts are more prevalent in surface waters when heavy rains increase runoff of wild and domestic animal wastes from the land, or when sewage treatment plants are overloaded or break down.

Only laboratories with specialized capabilities can detect the presence of *Cryptosporidium* oocysts in water. Unfortunately, present sampling and detection methods are unreliable. Recovering oocysts trapped on the material used to filter water samples is difficult. Once a sample is obtained, however, determining whether the oocyst is alive or whether it is the species *C. parvum* that can infect humans is easily accomplished by looking at the sample under a microscope.

The number of oocysts detected in raw (untreated) water varies with location, sampling time, and laboratory methods. Water treatment plants remove most, but not always all, oocysts. Low numbers of oocysts are sufficient to cause cryptosporidiosis, but the low numbers of oocysts sometimes present in drinking water are not considered cause for alarm in the general public.

Protecting water supplies from *Cryptosporidium* demands multiple barriers. Why? *Cryptosporidium* oocysts have tough walls that can withstand many environmental stresses and are resistant to the chemical disinfectants such as chlorine that are traditionally used in municipal drinking water systems and swimming pools.

Physical removal of particles, including oocysts, from water by filtration is an important step in the municipal water treatment process. Typically, water pumped from rivers or lakes into a treatment plant is mixed with coagulants (see Chapter 11), which help settle out particles suspended in the water. If sand filtration is used, even more particles are removed. Finally, the clarified water is disinfected and piped to customers. Filtration is the only conventional method now in use in the United States for controlling *Cryptosporidium*.

Ozone is a strong disinfectant (see Chapter 11) that kills protozoa if sufficient doses and contact times are used, but ozone leaves no residual for killing microorganisms in the distribution system as does chlorine. The high costs of new filtration or ozone treatment plants must be weighed against the benefits of additional treatment. Even well-operated water treatment plants cannot ensure that drinking water will be completely free of *Cryptosporidium* oocysts. Water treatment methods alone cannot solve the problem; watershed protection and monitoring of water quality are critical.

Watershed protection is another barrier to *Cryptosporidium* in drinking water. Land use controls such as septic systems regulations and best management practices to control runoff can help keep human and animal wastes out of water.

Under the Surface Water Treatment Rule of 1989, public water systems must filter surface water sources unless water quality and disinfection requirements are met and a watershed control program is maintained. This rule, however, did not address *Cryptosporidium*. The USEPA has now set standards for turbidity (cloudiness) and coliform bacteria (which indicate that pathogens are probably present) in drinking water. Frequent monitoring must occur to provide officials with early warning of potential problems to enable them to take steps to protect public health. Unfortunately, no water quality indicators can reliably predict the occurrence of cryptosporidiosis. More accurate and rapid assays of oocysts will make it possible to notify residents promptly if their water supply is contaminated with *Cryptosporidium* and thus avert outbreaks.

As Fayer et al. (1997) pointed out, efforts are underway to answer questions about the occurrence, detection, and treatment of *Cryptosporidium* in water so that the USEPA and states can set specific standards for this parasite in the future. The collaborative efforts of water utilities, government agencies, health care providers, and individuals are needed to prevent outbreaks of cryptosporidiosis.

Sidebar: Sydney, Australia[8]

From the end of July to the end of September 1998, upon three occasions, residents of the city of Sydney, Australia, had to take the precaution of boiling their drinking water. Testing found *Giardia* and *Cryptosporidium* in the public water supply.

According to the Sydney authorities, at these levels the *Giardia* and *Cryptosporidium* cysts posed little, if any, health threat. No incidents of illness were linked to the presence of *Giardia* and *Cryptosporidium*; however, businesses that rely on large quantities of pure water could not function on boil-alert quality water.

Evidence seemed to indicate that the plant itself was creating the problem, and that the results of water tests performed by the lab (Australian Water Technologies) were improperly read (or performed) and misinterpreted.

The aftermath of the incident has left the Sydney Water Corporation (the privately owned organization handling the treatment systems since privatization in January of 1995) in shambles. Beginning on the 29th of July, the three boil alerts (the last ended September 19, 1998) resulted in a massive investigation into the causes and sources of the contamination and the resignation of Sydney Water's managing director and the chairman (whose blossoming political career is in evident ruination). Sydney Water Corporation was also stripped of responsibility and major assets, losing control of treatment plants, dams, and catchment to the government's new Sydney Catchment Authority. Sydney Water Corporation must make repayments to residential water users for the expense and trouble of using bottled water, and a large lawsuit has begun, from the businesses and industries affected by the shutdown.

On the positive side, Australia is putting an American-style clean water act into place (which, if nothing else, will establish guidelines to follow in such a case), and Sydney is working to ensure good preventive maintenance measures for watershed protection.

Of special interest to water pollution control technologists are the tests, their results, and the difficulty in pinpointing the source (or sources) and cause of the contamination. U.S. experts warned that actually finding a direct source is unlikely (causes for the 1995 outbreak in Milwaukee are similarly uncertain), and recommended either an ozonation system or microfiltration be installed to ensure completely safe drinking water. That expert advice, though, presumes actual *Giardia* and *Cryptosporidium* contamination, which at this point is more than a little doubtful.

During that two-month period, test results on the same water samples

[8]Adapted from F. R. Spellman and N. E. Whiting, *Water Pollution Control Technology*, Rockville, MD: Government Institutes, 1999.

varied widely. Tests for the later shutdowns were less accurate than ones for the initial shutdown—not surprising given the panic conditions at the lab, under pressure to find the causes and gun-shy at the thought of risking either consumer wrath at inconvenience or consumer illness and death from contamination.

One test sample was read at 1000 *Cryptosporidium* oocysts initially. Sample re-examination found only two. Technicians may also have mistaken harmless algaes similar in appearance to *Giardia* and *Cryptosporidium* for the dangerous cysts, raising false (and expensive) alarms. Even in retesting, test results were shaky. The New South Wales Health Department counted what they thought were more than 9000 oocysts per hundred liters of treated water for one sample—higher levels than testing should find in raw sewage. An expert from the department of civil engineering at the University of New South Wales who saw lots of different algaes and no *Cryptosporidium* in his own tests on the water pointed out that the highest U.S. level ever reported was 1000, and that those U.S. reports included Milwaukee, 1995, where hundreds of thousands of people fell ill. He also pointed out that other common fecal bacteria should be present in the sample as well, but were not (*The Age,* 9/29/98).

The whole story in telegraph form comes forth in the headlines from Sydney's local press, beginning at the end of July, from the initial response to the first boil alert to the rising tide of legal action in January of 1999.[9]

Polluted water crisis—*The Age* (07/30/98)
Carr demands inquiry into Sydney water bug—AGE(07/30/98)
Message not reaching some communities—*Sydney Morning Herald* (07/30/98)
Water bug a threat to AIDS sufferers—AGE (07/30/98)
Independent inquiry announced—AGE (07/30/98)
Sydney's water bug: Please explain—AGE (07/30/98)
Sydney water scare—AGE (07/30/98)
Water crisis extends into weekend—Australian Broadcasting Corporation (07/31/98)
"Bathwater" slakes the city's thirst—SMH (07/31/98)
Dam rangers cut to a trickle—The Australian News Network (07/31/98)
All of Sydney under water contamination alert—ABC (07/31/98)
No, it's not Calcutta—it's Sydney—AGE (07/31/98)
How Sydney was kept in the dark—AGE (07/31/98)
Carr orders water enquiry—ANN (07/31/98)
Huge operation to flush the system—ANN (07/31/98)
How we joined the Third World—ANN (07/31/98)
Water crisis grips Sydney—SMH (07/31/98)

---

[9]For access to the actual articles, go to http://headlines.yahoo.com/Full_Coverage/aunz/polluted_water_in_sydney/.

A taste of the Third World—SMH (07/31/98)
No need for panic, say doctors—SMH (07/31/98)
Do panic! There's a bug in the system—SMH (07/31/98)
Kids learn how to beat the "germs"—SMH (07/31/98)
Avoid tap water from now on, says AIDS Council—SMH (07/31/98)
Employers advised to bring in the bottles—SMH (07/31/98)
Dogs not to blame—ANN (08/01/98)
Jokes about water turn to whine—AGE (08/01/98)
Source of tainting a challenge to trace—SMH (08/01/98)
Six-year warning on parasites—SMH (08/01/98)
Sydney's down-under solution—SMH (08/01/98)
Honesty deserves respect, industry leader claims—SMH (08/01/98)
Broken public trust will take years to repair—SMH (08/01/98)
Environment groups warn of wider health threat—SMH (08/01/98)
Water supplier may pay dearly—SMH (08/01/98)
Minute parasites live in the gut—SMH (08/01/98)
Test results remain secret—SMH (08/01/98)
Pooling its problem—SMH (08/01/98)
Restricted warning a blunder: minister—ANN (08/01/98)
Water bug could surface anywhere—ANN (08/01/98)
Our bottlers heed call to spring into action—ANN (08/01/98)
Water's usually a pure wonder—ANN (08/01/98)
Minister pledges to find culprit—AGE (08/01/98)
Here's to Sydney Water—SMH (08/01/98)
Melbourne's water is a very good catch—AGE (08/02/98)
Water all clear expected for some Sydney suburbs—ABC (08/02/98)
Tap water safe for some, but testing continues—AGE (08/03/98)
Experts tell Sydney to invest in $100m filter—AGE (08/03/98)
Bungled from day one—SMH (08/04/98)
Water inquiry head fired—ANN (08/04/98)
Safe water suburbs—ANN (08/04/98)
Minister widens net on scandal—ANN (08/04/98)
Lawyers the big winners in crisis—ANN (08/04/98)
The cash that got away—ANN (08/04/98)
Business counts costs and options—SMH (08/04/98)
Troubles always on tap for Chairman Hill—SMH (08/04/98)
Claims Sydney Water's actions indicate lack of catchment care—ABC (08/04/98)
Fit to drink, now Carr targets water culprits—SMH (08/05/98)
You can lead a chairman to water . . .—SMH (08/05/98)
Remedy may cost $300m—SMH (08/05/98)
Not a drama, down in cappuccino country—SMH (08/05/98)
Catchments to be sewered—SMH (08/05/98)
Water bugs predated Prospect—SMH (08/05/98)
Safe water: the big lie—SMH (08/07/98)
Invasion of parasites raises doubts over plant—SMH (08/07/98)
A creek that tests like a sewer—SMH (08/07/98)

Opposition demands Hill's scalp over "interference"—SMH (08/07/98)
The cash has already been spent—SMH (08/07/98)
Catchment areas better protected under new plans—SMH (08/07/98)
Water bug runs to form after crisis—SMH (08/11/98)
Hill won't resign over water contamination—ABC (08/20/98)
Results of water test: chief resigns—SMH (08/20/98)
Genuine health risk, says inquiry chief—SMH (08/20/98)
Hill jumps from water crisis role—SMH (08/21/98)
Treatment plant may be liable for foul water crisis—SMH (08/21/98)
Troubleshooter Hill a real brick—SMH (08/21/98)
New warning: boil all water—SMH (08/26/98)
Health department re-issues water warning for Sydney—ABC (08/26/98)
Good results for latest water testing in Sydney—ABC (08/27/98)
End of an era for reliable, safe water—SMH (08/27/98)
Tougher law needed, says ACA—SMH (08/27/98)
After the blame, only a cure matters—SMH (08/27/98)
High readings just kept coming in—SMH (08/27/98)
Legal case to involve up to 500 claimants—SMH (08/27/98)
Why our fishmongers are already fed up to the gills—SMH (08/27/98)
Bugs fail to advance on the waterfront—SMH (08/27/98)
Bugs found in Sydney's main water source—ABC (08/28/98)
Water alert to remain over weekend—SMH (08/28/98)
The Collins solution: add new expertise and stir—SMH (08/28/98)
Blooming algae, not bugs, blamed for crisis—SMH (08/29/98)
Water on the mend but cause still unknown—SMH (08/31/98)
City to start going off the boil from today—SMH (09/01/98)
Sydney Water braces for court action—SMH (09/02/98)
Federal offer on Sydney water—AGE (09/08/98)
Ozone could plug hole in water quality—SMH (09/08/98)
Three in four are heeding official advice—SMH (09/08/98)
Scare just water off a tourist's back—SMH (09/08/98)
Bugs spread in sludge in catchment areas—SMH (09/08/98)
Latest health threat: apathy—SMH (09/08/98)
Minister's "grubby politics"—SMH (09/08/98)
Water bug could be here to stay—ANN (09/09/98)
Water authority plays it safe on danger bugs—AGE (09/09/98)
Biggest overseas turn-off since Hanson, says Stone—ANN (09/09/98)
Parliament in uproar over water fallout—SMH (09/09/98)
Blueprint to keep water safe is put on hold—SMH (09/10/98)
Home filters guaranteed to get (nearly) all the bugs out—SMH (09/10/98)
Prospect mystery—SMH (09/10/98)
IOC to grill Sydney over water safety—AGE (09/11/98)
Games chiefs seek water assurance—SMH (09/11/98)
"Higher-quality" water to improve untreated supply—SMH (09/11/98)
Warned to be on giardia—SMH (09/11/98)
Pool owners advised to add extra chlorine—SMH (09/11/98)
Sydney gets water in its chlorine supply—ANN (09/11/98)

Water bugs ignored—ANN (09/13/98)
Water all-clear in doubt—ANN (09/14/98)
High contamination found 8 months ago—SMH (09/15/98)
It never rains, it pours for bottled water industry—SMH (09/15/98)
Radio daze could buzz off those nasty bugs—SMH (09/15/98)
Water may be unsafe for 2 years—ANN (09/16/98)
HIV-positive man to sue Sydney Water—SMH (09/16/98)
Water off line after Prospect shutdown—SMH (09/16/98)
Water back on—no guarantees—ANN (09/20/98)
Water chiefs take leave—ANN (09/23/98)
MPs demand all documents on water crisis—SMH (09/25/98)
Sydney water scare back on boil—AGE (09/27/98)
Water bugs are algae, says expert—AGE (09/29/98)
Water inquiry given more power—SMH (10/22/98)
McLellan bid to dilute power of Sydney Water—SMH (10/29/98)
Water: now the payback—SMH (10/30/98)
McClellan leaves water hierarchy shaken and stirred—SMH (10/30/98)
Cave-in on water to cost millions—SMH (11/04/98)
A fresh start for water—SMH (12/04/98)
Water crisis linked to ignorance factor—SMH (12/16/98)
Inquiry outcome dribbles out—SMH (12/16/98)
Dumping spoil at sea "final option"—SMH (01/09/99)
Polluted tap-water claimants line up—SMH (01/13/99)

The moral of the story? Know your stuff. Be sure of your technique and do everything you can to ensure the accuracy of your samples and test results.

As a water control technologist, you may find you have to turn in test results that open a similar can of worms at some point, or announce unsafe water supplies to the press.

6.5.1.2.2 Cryptosporidiosis

Dennis D. Juranek (1995) from the Centers for Disease Control has written in *Clinical Infectious Diseases:*

Cryptosporidium parvum is an important emerging pathogen in the United States and a cause of severe, life-threatening disease in patients with AIDS. No safe and effective form of specific treatment for cryptosporidiosis has been identified to date. The parasite is transmitted by ingestion of oocysts excreted in the feces of infected humans or animals. The infection can therefore be transmitted from person-to-person, through ingestion of contaminated water (drinking water and water used for recreational purposes) or food, from animal to person, or by contact with fecally contaminated environmental surfaces. Outbreaks associated with all of these modes of transmission have been documented. Patients with human immunodeficiency virus infection should be made more aware of the many ways that *Cryptosporidium* species are transmitted, and they should be given guidance on how to reduce their risk of exposure.

The CDC (1995) points out that since the Milwaukee outbreak, concern about the safety of drinking water in the U.S. has increased, and new attention has been focused on determining and reducing the risk of cryptosporidiosis from community and municipal water supplies.

Cryptosporidiosis is spread by putting something in the mouth that has been contaminated with the stool of an infected person or animal. In this way, people swallow the *Cryptosporidium* parasite. As we pointed out earlier, a person can become infected by drinking contaminated water or eating raw or undercooked food contaminated with *Cryptosporidium* oocysts; direct contact with the droppings of infected animals or stools of infected humans; or hand-to-mouth transfer of oocysts from surfaces that may have become contaminated with microscopic amounts of stool from an infected person or animal.

The symptoms may appear two to ten days after infection by the parasite. Although some persons may not have symptoms, others have watery diarrhea, headache, abdominal cramps, nausea, vomiting, and low-grade fever. These symptoms may lead to weight loss and dehydration.

In otherwise healthy persons, these symptoms usually last one to two weeks, at which time the immune system is able to stop the infection. In persons with suppressed immune systems, such as persons who have AIDS or who recently have had an organ or bone marrow transplant, the infection may continue and become life-threatening.

At the present time, no safe and effective cure for cryptosporidiosis exists. People with normal immune systems improve without taking antibiotic or antiparasitic medications. The treatment recommended for this diarrheal illness is to drink plenty of fluids and to get extra rest. Physicians may prescribe medication to slow the diarrhea during recovery.

The best way to prevent cryptosporidiosis is

- Avoid water or food that may be contaminated.
- Wash hands after using the toilet and before handling food.
- If you work in a child care center where you change diapers, be sure to wash your hands thoroughly with plenty of soap and warm water after every diaper change, even if you wear gloves.

During community-wide outbreaks caused by contaminated drinking water, drinking water practitioners should inform the public to boil drinking water for one minute to kill the *Cryptosporidium* parasite.

6.5.1.3 Cyclospora

Cyclospora organisms, which until recently were considered blue-green algae, were discovered at the turn of the century. The first human cases of *Cyclospora* infection were reported in the 1970s. In the early 1980s, *Cyclospora* was recognized as a pathogen in patients with AIDS. We now know that *Cyclospora* is endemic in many parts of the world, and appears to be an important cause of traveler's diarrhea. *Cyclospora* are two to three times larger than *Cryptosporidium*, but otherwise have similar features. *Cyclospora* diarrheal illness in patients with healthy immune systems can be cured with a week of therapy with timethoprim-sulfamethoxazole (TMP-SMX).

Huang et al. (1995) detailed what they believe is the first known outbreak of diarrheal illness associated with *Cyclospora* in the United States. The outbreak, which occurred in 1990, consisted of 21 cases of illness among physicians and others working at a Chicago hospital. Contaminated tap water from a physicians' dormitory at the hospital was the probable source of the organism. The tap water probably picked up the organism while in a storage tank at the top of the dormitory after the failure of a water pump.

Watery diarrhea, abdominal cramping, low-grade fever, and decreased appetite are common features of the illness, investigators reported in *Annals of Internal Medicine*. The illness also is marked by periods of remission and relapse that may continue for up to several weeks.

Microscopic examination of stool specimens from 11 infected people showed many spherical bodies 8 to 10 µm in diameter that were identified as a *Cyclospora* species. The only other outbreaks associated with *Cyclospora* in the literature have been seasonal outbreaks in Nepal. One outbreak in Nepal was associated with chlorinated drinking water.

6.5.1.3.1 The Basics of Cyclospora

In 1998, the CDC described *Cyclospora cayetanensis* as a unicellular parasite previously known as a cyanobacterium-like (blue-green algae-like) or coccidia-like body (CLB). Since the first known cases of illness caused by *Cyclospora* infection were reported in the medical journals in the 1970s, cases have been reported with increased frequency from various countries since the mid-1980s (in part because of the availability of better techniques for detecting the parasite in stool specimens).

The transmission of *Cyclospora* is not a straightforward process. When infected persons excrete the oocyst state of *Cyclospora* in their feces, the oocysts are not infectious and may require from days to weeks to become so (i.e., to sporulate). Therefore, transmission of *Cyclospora* directly from an infected person to someone else is unlikely. However, indirect transmission can occur if an infected person contaminates the environment and oocysts have sufficient time, under appropriate conditions, to become infectious. For example, *Cyclospora* may be transmitted by ingestion of water or food contaminated with oocysts. Outbreaks linked to contaminated water, as well as outbreaks linked to various types of fresh produce, have been reported in recent years (Herwaldt et al., 1997; CDC, 1997a; CDC, 1997b; and Huang et al., 1995). How common the various modes of transmission and sources of infection are is not yet known, nor is it known whether animals can be infected and serve as sources of infection for humans.

Persons of all ages are at risk for infection. Persons living or traveling in developing countries may be at increased risk; but infection can be acquired worldwide, including in the United States. In some countries of the world, infection appears to be seasonal.

The incubation period between acquisition of infection and onset of symptoms averages one week. *Cyclospora* infects the small intestine and typically causes watery diarrhea, with frequent, sometimes explosive, stools. Other symptoms can include loss of appetite, substantial loss of weight, bloating, increased flatus, stomach cramps, nausea, vomiting, muscle aches, low-grade fever, and fatigue. If untreated, illness may last for a few days to a month or longer, and may follow a remitting-relapsing course. Some infected persons are asymptomatic.

Identification of this parasite in stool requires special, not routine, laboratory tests be done. A single negative stool specimen does not rule out the diagnosis; three or more specimens may be required. Stool specimens should also be checked for other microbes that can cause a similar illness.

In providing treatment, TMP-SMX has been shown in a placebo-controlled trial to be effective treatment for *Cyclospora* infection (Hoge et al., 1995). No alternate antibiotic regimen has been identified yet for patients who do not respond to or are intolerant of TMP/SMX.

Based on currently available information, avoiding food or water that may be contaminated with stool is the best way to prevent infection. Reinfection can occur.

Huang et al. (1995) list the following key points for laboratory diagnosis of *Cyclospora*:

(1) To maximize recovery of *Cyclospora* oocysts, first concentrate the stool specimen by the formalin-ethyl acetate technique (centrifuge for 10 minutes at $500 \times g$) and then examine a wet mount and/or stained slide of the sediment.

(2) *Cyclospora* oocysts are 8–10 microns in diameter (in contrast, *Cryptosporidium parvum* oocysts are 4–6 microns in diameter).

(3) Ultraviolet epifluorescence microscopy is a sensitive technique for

rapidly examining stool sediments for *Cyclospora* oocysts, which auto-fluoresce (*Cryptosporidium parvum* oocysts do not). If suspect oocysts are found, bright-field microscopy can then be used to confirm that the structures have the characteristic morphologic features of *Cyclospora* oocysts (i.e., are nonrefractile spheres that contain undifferentiated cytoplasm of refractile globules).

(4) On a modified acid fast-stained slide of stool (the technique used by most laboratories), *Cyclospora* oocysts are variably acid fast (i.e., in the same field, oocysts may be unstained or stained from light pink to deep red). Unstained oocysts may have a wrinkled appearance; observers must distinguish oocysts from artifacts that may be acid fast but do not have the all-important wrinkled morphology of the oocyst wall.

(5) Using a modified safranin technique, oocysts uniformly stain a brilliant reddish orange if fecal smears are heated in a microwave oven during staining (Visvesvara et al., 1997). If epifluorescence microscopy is available, the stained slide can first be examined with this technique and suspect oocysts re-examined with bright-field microscopy.

(6) Although not recommended as an optimal technique for detection of *Cyclospora*, on a trichrome-stained slide of stool, the oocysts appear as clear, round, and somewhat wrinkled.

6.6 HELMINTHS

The National Academy of Sciences (NAS, 1977) considers that drinking water in the United States may transmit the following intestinal worms (nematodes):

- *Ascaris lumbricoides*—stomach worm
- *Trichuris trichiura*—whip worm
- *Ancylostoma duodenale*—hookworm
- *Necator americanus*—hookworm
- *Strongyloides stercoralis*—threadworm

Along with inhabiting organic muds, worms also inhabit biological slimes and are derived from sewage and wet soil. Nematodes multiply in wastewater treatment plants; strict aerobes, they have been found in activated sludge and particularly in trickling filters, and therefore appear in large concentration in treated domestic liquid waste. Microscopic in size, they range in length from 0.5–3 mm and in diameter from 0.01–0.05 mm. Most species have a similar appearance. They have a body that is covered by cuticle, are cylindrical, nonsegmented, and taper at both ends.

Nematodes ingest bacteria, and entering the distribution system may protect pathogens from disinfection of the water supply and reach the consumer.

According to NAS (1977), active motile nematode larvae can penetrate sand filters and survive chlorination, but they are not normally expected to cause parasitic nematode infections.

Free-living nematodes have a life cycle consisting of egg, four larval stages, and one adult stage. Eggs are easily recognizable in finished water, while raw water must have excessive microfaunal forms to allow identification.

Environmental conditions have an impact on the growth of nematodes. For example, in anoxic conditions, their swimming and growth is impaired. Temperature fluctuations directly affect their growth and survival; population decreases when temperatures increase.

Aquatic flatworms (improperly named because they are not all flat) feed primarily on algae. Because of their aversion to light, they are found in the lower depths of pools. Flatworms are very hardy and can survive in wide variations in humidity and temperature.

Surface waters that are grossly polluted with organic matter (especially domestic sewage) have a fauna capable of thriving in very low concentrations of oxygen. A few species of tubificid worms dominate this environment. Pennak (1989) reports that the bottoms of severely polluted streams can be literally covered with a "writhing mass" of these tubificids.

For the drinking water practitioner interested in learning more about aquatic worms, *Standard Methods* (20th Edition, 1999) has a section for *Nematological Examination*, detailing sample collection and printing an "Illustrated Key to Fresh Water Nematodes," listing more than 100 names and graphical representations.

6.7 SUMMARY

De Zuane (1997) points out that pathogenic parasites are not easily removed or eliminated completely by conventional treatment and disinfection unit processes. This is particularly true for *Giardia lamblia, Cryptosporidium,* and *Cyclospora.* Filtration facilities can be adjusted in depth, prechlorination, filtration rate, and backwashing to become more effective in the removal of cysts. The pretreatment of protected watershed raw water is a major factor in the elimination of pathogenic protozoa (see Chapter 11).

6.8 REFERENCES

The Age. "Water Bugs Are Algae, Says Expert." Sydney, Australia, September 29, 1998.

Badenock, J., Cryptosporidium *in Water Supplies.* London: HMSO Publications, 1990.

Bergey's Manual of Determinative Bacteriology, 9th Edition, ed., J. G. Holt, Philadelphia: Williams & Wilkins, 1993.

Bingham, A. K., Jarroll, E. L., Meyer, E. A., and Radulescu, S., *Introduction of Giardia Excystation and the Effect of Temperature on Cyst Viability Compared by Eosin-Exclusion and in vitro Excystation in Waterborne Transmission of Giardiasis*, ed., J. Jakubowski and H. C. Hoff. Washington, D.C.: United States Environmental Protection Agency, pp. 217–229, EPA-600/9-79-001, 1979.

Black, R. E., Dykes, A. C., Anderson, K. E., Wells, J. G., Sinclair, S. P., Gary, G. W., Hatch, M. H., and Gnagarosa, E. J., "Handwashing to Prevent Diarrhea in Day-Care Centers." *Am. J. Epidemiol.* 113:445–451, 1981.

Black-Covilli, L. L., "Basic Environmental Chemistry of Hazardous and Solid Wastes." In *Fundamentals of Environmental Science and Technology*, ed., P. C. Knowles. Rockville, Maryland: Government Institutes, Inc., pp. 13–30, 1992.

Brodsky, R. E., Spencer, H. C., and Schultz, M. G., "Giardiasis in American Travelers to the Soviet Union." *J. Infect. Dis.* 130:319–323, 1974.

Carson, R., *Lost Woods: The Discovered Writing of Rachel Carson*, ed., L. Lear. Boston: Beacon Press, 1988.

CDC. *Intestinal Parasite Surveillance, Annual Summary 1978*. Atlanta: Centers for Disease Control, 1979.

CDC. *Water-Related Disease Outbreaks Surveillance, Annual Summary 1983*. Atlanta: Centers for Disease Control, 1984.

CDC. *Cryptosporidiosis (Fact Sheet)*. Atlanta: Centers for Disease Control, 1995.

CDC. *Update: Outbreaks of Cyclosporiasis—United States and Canada.* MMWR 46:521–523, 1997a.

CDC. *Outbreak of Cyclosporiasis*—northern Virginia-Washington, D.C.-Baltimore, Maryland, Metropolitan Area. MMWR 46:689–691, 1997b.

Craun, G. F., *Waterborne Outbreaks of Giardiasis—Current Status in Giardia and Giardiasis*, ed., S. L. Erlandsen and E. A. Meyer. New York: Plenum Press, pp. 243–261, 1984.

Davidson, R. A., "Issues in Clinical Parasitology: The Treatment of Giardiasis." *Am J. Gastroenterol.* 79:256–261, 1984.

De Zuane, J., *Handbook of Drinking Water Quality*. New York: John Wiley and Sons, Inc., 1997.

Fayer, R., Speer, C. A., and Dubey, J. P., "The General Biology of *Cryptosporidium*." In Cryptosporidium *and Cryptosporidiosis*, ed., Ronald Fayer. Boca Raton, Florida: CRC Press, 1997.

Frost, F., Plan, B., and Liechty, B., "*Giardia* Prevalence in Commercially Trapped Mammals." *J. Environ. Health* 42:245–249, 1984.

Herwaldt, B. L., et al., "An Outbreak in 1996 of Cyclosporiasis Associated with Imported Raspberries." *N. Engl. J. Med.* 336:1548–1556, 1997.

Hoge, C. W., et al., "Placebo-Controlled Trail of Co-Trimoxazole for Cyclospora Infections among Travelers and Foreign Residents in Nepal." *Lancet* 345:691–693, 1995.

Huang, P., Weber, J. T., Sosin, D. M., et al., "Cyclospora." *Ann. Intern Med.* 123:401–414, 1995.

Jarroll, E. L., Jr., Bingham, A. K., and Meyer, E. A., "*Giardia* Cyst Destruction: Effectiveness of Six Small-Quantity Water Disinfection Methods." *Am. J. Trop. Med. Hygiene* 29:8–11, 1979.

Jarroll, E. L., Jr., Bingham, A. K., and Meyer, E. A., "Inability of an Iodination Method to Destroy Completely *Giardia* Cysts in Cold Water." *West J. Med.* 132:567–569, 1980.

Jarroll, E. L., Bingham, A. K., and Meyer, E. A., "Effect of Chlorine on *Giardia lamblia* Cyst Viability." *Appl. Environ. Microbiol.* 41:483–487, 1981.

Jokipii, L., and Jokipii, A. M. M.," Giardiasis in Travelers: A Prospective Study." *J. Infect. Dis.* 130:295–299, 1974.

Juranek, D. D., *"Cryptosporidium parvum."* *Clinical Infectious Diseases.* Atlanta: 21 (Suppl. 1) S57–61, 1995.

Juranek, D. D., *Giardiasis.* U.S. Centers for Disease Control, 1995.

Keystone, J. S., Karden, S., and Warren, M. R., "Person-to-Person Transmission of *Giardia lamblia* in Day-Care Nurseries." *Can. Med. Assoc. J.* 119:241–242, 247–248, 1978.

Keystone, J. S., Yang, J., Grisdale, D., Harrington, M., Pillow, L., and Andrychuk, R., "Intestinal Parasites in Metropolitan Toronto Day-Care Centres." *Can. Med. Assoc. J.* 131:733–735, 1984.

Kordon, C., *The Language of the Cell.* New York: McGraw-Hill, Inc., 1993.

Koren, H., *Handbook of Environmental Health and Safety: Principles and Practices.* Chelsea, MI: Lewis Publishers, 1991.

LeChevallier, M. W., Norton, W. D, and Lee, R. G., "Occurrence of *Giardia* and *Cryptosporidium* spp. in Surface Water Supplies." *Applied and Environmental Microbiology* 57(9):2610–2616, 1991.

Marchin, B. L., Fina, L. R., Lambert, J. L., and Fina, G. T., "Effect of Resin Disinfectants—13 and 15 on *Giardia muris* and *Giardia lamblia.* *Appl. Environ. Microbiol.* 46:965–969, 1983.

Mayo Foundation. *The "Bug" That Made Milwaukee Famous.* Mayo Foundation for Medical Education and Research, 1996.

National Academy of Sciences. *Drinking Water and Health*, (Vol. 1). Washington, D.C.: The National Research Council, National Academy Press, 1977.

National Academy of Sciences. *Drinking Water and Health*, (Vol. 4). Washington, D.C.: The National Research Council, National Academy Press, 1982.

Panciera, R. J., Thomassen, R. W., and Garner, R. M., "Cryptosporidial Infection in a Calf." *Vet. Pathol.* 8:479, 1971.

Patterson, D. J. and Hedley, S., *Free-Living Freshwater Protozoa: A Color Guide.* Boca Raton, Florida: CRC Press, Inc., 1992.

Pickering, L. K., Evans, D. G., Dupont, H. L., Vollet, J. J., III, and Evans D. J., Jr., "Diarrhea Caused by *Shigella,* Rotavirus, and Giardia in Day-Care Centers: Prospective Study." *J. Pediatr.*, 99:51–56, 1981.

Pickering, L. K., Woodward, W. E., Dupont, H. L., and Sullivan, P., "Occurrence of *Giardia lamblia* in Children in Day Care Centers." *J. Pediatr.* 104:522–526, 1984.

Pennack, R. W., *Freshwater Invertebrates of the United Stated,* 3rd Edition. New York: John Wiley and Sons, p. 24, 1989.

Prescott, L. M., Harley, J. P., and Klein, D. A., *Microbiology,* 3rd Edition. Dubuque, Iowa: Wm. C. Brown Company Publishers, 1993.

Rendtorff, R. C., "The Experimental Transmission of Human Intestinal Protozoan Parasites. II. *Giardia lamblia* Cysts Given in Capsules." *Am. J. Hygiene* 59:209–220, 1954.

Sealy, D. P. and Schuman, S. H., "Endemic Giardiasis and Day Care." *Pediatrics* 72:154–158, 1983.

Singleton, P., *Introduction to Bacteria*, 2nd Edition. New York: John Wiley and Sons, 1992.

Singleton, P. and Sainsbury, D., *Dictionary of Microbiology and Molecular Biology*, 2nd Edition. New York: John Wiley and Sons, 1994.

Slavin, D., "*Cryptosporidium meleagridis* (s. nov.)." *J. Comp. Pathol.* 65:262, 1955.

Spellman, F. R., *Microbiology for Water/Wastewater Operators*. Lancaster, PA: Technomic Publishing Company, Inc., 1997.

Spellman, F. R. and Whiting, N. E., *Water Pollution Control Technology*. Rockville, MD: Government Institutes, 1999.

Standard Methods for Examination of Water and Wastewater, 20th Edition, eds., A. E. Greenberg et at. American Public Health Assn., 1999.

Tchobanoglous, G. and Schroeder, E. D., *Water Quality*. Reading, Massachusetts: Addison-Wesley Publishing Company, 1987.

Thomas, L., *The Lives of a Cell*. New York: Viking Press, 1974.

Thomas, L., *Late Night Thoughts on Listening to Mahler's Ninth Symphony*. New York: Viking Press, 1982.

Tyzzer, E. E., "A Sporozoan Found in the Peptic Glands of the Common Mouse." *Proc. Soc. Exp. Biol. Med.* 1907.

Tyzzer, E. E., "*Cryptosporidium parvum* (sp. nov.), a Coccidium Found in the Small Intestine of the Common Mouse." *Arch. Protistenkd,* 26:394, 1912.

Upton, S. J., *Basic Biology of* Cryptosporidium. Kansas State University, 1997.

Visvesvara, G. S., et al., "Uniform Staining of *Cyclospora* Oocysts in Fecal Smears by a Modified Safranin Technique with Microwave Heating." *J. Cln. Microbiol* 35:730–733, 1997.

Walsh, J. A., Estimating the Burden of Illness in the Tropics. In *Tropical and Geographic Medicine*, ed., K. S. Warren and A. F. Mahmoud. New York: McGraw-Hill, pp. 1073–1085, 1981.

Walsh, J.D. and Warren K. S., "Selective Primary Health Care: An Interim Strategy for Disease Control in Developing Countries." *N. Engl. J. Med.,* 301:974–976, 1979.

Weller, P. F., "Intestinal Protozoa: Giardiasis." *Scientific American Medicine.* 1985.

WHO. *Guidelines for Drinking Water Quality. (Vol. 1—Recommendation; Vol. 2—Health Criteria and Other Supporting Information)*. Geneva, Switzerland: World Health Organization, 1984.

Wistreich, G. A. and Lechtman, M. D., *Microbiology*, 3rd Edition. New York: Macmillan Publishing Co., 1980.

CHAPTER 7

Drinking Water Parameters:
Physical

Water is H₂O, hydrogen two parts, oxygen one, but there is also a third thing,
that makes it water and nobody knows what that is. (D. H. Lawrence, Pansies,
"The Third Thing," 1929)

7.1 INTRODUCTION

WHAT would you say if, at a given moment, you were asked to describe at least one physical property of water? Our most likely response would be, "Water is wet." That is not only the simplest response, but also the one that is most obvious to us. Water is indeed wet. Water's wetness is perhaps its most obvious physical characteristic.

But let's assume the query did not end there. Let's assume that the next question posed to us is, "What makes water wet?" With this question, we leave the realm of general knowledge; and unless we are practitioners of water science, this question may be beyond our ability to answer.

According to David Clary (1997), a chemist at University College London, water does not start to behave like a liquid until at least six molecules form a cluster. He found that groups of five water molecules or fewer have planar structures, forming films one molecule thick. However, when a sixth molecule is added, the cluster switches to a three-dimensional cage-like structure and suddenly it has the properties of water—it becomes wet.

Other physical characteristics of water that we are interested in are somewhat more germane to the discussion at hand; namely, categories and parameters to be used to define the physical water quality of a particular water supply (any water supply). One such category includes the physical characteristics of water detectable by the senses of smell, taste, sight, and touch. Taste and odor, color, temperature, turbidity, and solids fall into this category (see Figure 7.1). These physical characteristics are rather obvious, too; maybe not as obvious as water's being wet, but just as important.

Because we discuss in this chapter the physical characteristics that apply

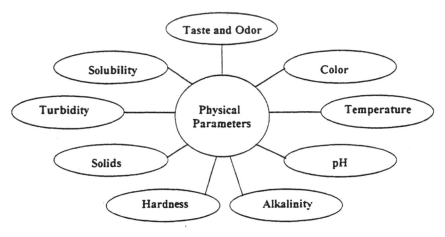

Figure 7.1 Physical or general water parameters.

both to principles of drinking water evaluation and analysis of water found in nature as potential or actual sources of water supplies, we must include additional physical characteristics of water that are not so obvious.

In Figure 7.1, you see that in the sections that follow, we include the traditional physical parameters of taste and odor, color, temperature, turbidity, and solids. Figure 7.1 also includes pH and shows that we examine solids. We compare alkalinity and hardness, though they can also be considered chemical parameters. Lastly, because of the importance of pH in water treatment, we review the basic concepts of water solubility.

Before we begin a discussion of physical water quality parameters, we point out that when this text refers to "water quality" our definition is predicated on the intended use of the water (in our case, for potable use). Over the years many parameters have evolved that qualitatively reflect the impact that various contaminants (impurities) have on water used for potable purposes; the following sections provide a brief discussion of these parameters.

7.2 TASTE AND ODOR

Note: Remember that under the Safe Drinking Water Act (SDWA), the USEPA issued guidelines to the states regarding secondary drinking water standards. These guidelines apply to drinking water contaminants that may adversely affect the aesthetic qualities of water such as odor and appearance. These qualities have no known adverse health effects, and thus secondary regulations are not mandatory. However, most drinking water systems comply with the limits; they have learned through experience that though the odor and appearance of drinking water is not a problem until the customer complains, one thing is certain—the customer does complain, and complain quite often (Spellman, 1998). Disinfection itself often becomes one of the major sources of complaint. For example, probably the most often heard complaint by drinking water consumers is "chlorine taste," where the odor threshold is sometimes as low as 0.2–0.4 mg/L at the typical pH level (De Zuane, 1997).

Taste and *odor* are used jointly in the vernacular of water science. As we said above, in drinking water, taste and odor are not normally a problem until the consumer complains; drinking water practitioners soon learn through experience that taste and odor problems may be their first alarm signal for a potential health hazard. Taste and odor are thus important for aesthetic reasons (as a measure of the acceptability of water), with little impact on how safe the water is to drink, but should not be ignored.

Objectionable taste and odor are more likely found at the raw water source than at the consumer's tap. In general, water contaminants are attributable to contact with nature or human use. Taste and odor in water are caused by a variety of substances, including minerals, metals and salts from the soil, constituents of wastewater, and end products produced in biological reactions. Earthy, musty odors are common in some water supplies and normally are derived from natural biological processes. The more offensive odors [those caused by hydrogen sulfide gas (H_2S), for example] are also common in water supplies. The rotten-egg smell of this gas may be encountered in water that has been in contact with naturally occurring deposits of decaying organic matter. Groundwater supplies sometimes have this problem; the wells are commonly called *sulfur wells*.

Problems with tastes and odors are normally associated with surface rather than groundwater. Surface water taste and odor problems are normally caused by algae and other microorganisms. Groundwater taste and odor problems are generally the result of human interference/influence, in particular, landfill leachate.

Note: Human ability to detect odor thresholds of various substances in water ranges from a low of about 1 mg/L for methylisoborneol to a high of about 20 mg/L for chloroform.

In testing performed in laboratories, consult *Standard Methods for the Examination of Water and Wastewater* (APHA-AWWA-WEF, 1995), Appendix A-VIII, "General References." The qualitative terms used to describe taste and odor are often classified as grassy, swampy, septic, musty, fishy, phenolic, and sweet.

In water treatment, one of the common methods used to remove taste and odor is to oxidize the materials that cause the problem with oxidants, including potassium permanganate and chlorine. Another common treatment method is to feed powdered activated carbon prior to the filter. The activated carbon has numerous small openings that adsorb the components that cause the odor and tastes. Taste and odor problems are also often controlled by watershed management, use of algicide, aeration, and pretreatment. Properly functioning water filtration systems help to minimize taste and odor problems as well.

Odor is typically measured and expressed in terms of a *threshold odor number* (TON), the ratio by which the sample has to be diluted for the odor to become virtually unnoticeable. The USEPA,[10] in 1989, issued a "Secondary Maximum Contaminant Level" (SMCL) of 3 TON for odor. (*Note:* Remember, secondary standards are parameters not related to health.)

When a dilution is used, a number can be devised in clarifying odor.

$$\text{TON} = \frac{V_T + V_D}{V_T} \qquad (7.1)$$

where:

V_T = volume tested
V_D = volume of dilution with odor-free distilled water

for $V_D = 0$, TON = 1 (lowest value possible); for $V_D = V_T$, TON = 2; for $V_D = 2V_T$, TON = 3, etc.

Note: Though taste and odor (along with color) are seldom connected to toxicological effects, the drinking water practitioner should never be fooled into assuming that a water supply with a "bit of" taste and odor will not offend the consumer—it will.

7.3 COLOR

The quality of water can also be judged by its color, and the consumer does so, at least from a psychological point of view. Go to the kitchen tap and draw a glass of water. What if, surprisingly, it were rust-colored, maybe to the point where you couldn't even see your fingers on the other side of the glass? In fact, the rust-colored water may be safe to drink, but will the average consumer think so?

Pure water is colorless, but water in nature is often colored by foreign substances, including organic matter from soils, vegetation, minerals, and aquatic organisms that are often present in natural waters. Color can also be contributed by municipal and industrial wastes.

Color in water is classified as either *true color* or *apparent color*. Color

---

[10]Refer to Part 143—National Secondary Drinking Water Regulations—*Federal Register,* Vol. 54, No. 97, May 22, 1989.

in water that is partly due to dissolved solids that remain after removal of suspended matter is known as true color. Color contributed by suspended matter is called apparent color. In water treatment, true color is the most difficult to remove. Color is measured by comparing the water sample with standard color solutions or colored glass disks. One *color unit* (CU) is equivalent to the color produced by a 1 mg/L solution of platinum. The USEPA, in 1989, as a guide issued a Secondary MCL of 15 color units for color. At 10–15 color units, color may not be visually detectable; at 100 color units, water may have the appearance of tea.

Note: In practice, isolating and identifying specific chemicals that cause the color is not practical.

As we've said, the impact of color in water is a matter of aesthetics; consumers do not find it acceptable. No matter how safe the water may be to drink, most people object strongly to water that offends their sense of sight. Given a choice, the public, obviously, would prefer clear, uncolored water.

The effects of water color, though, go beyond the psychological implications. Colored water affects laundering, papermaking, manufacturing, textiles, and food processing. The color of water has a profound effect on its marketability for both domestic and industrial use.

In water treatment, color is not usually considered unsafe or unsanitary, but it is a treatment problem because it exerts a chlorine demand, which reduces the effectiveness of chlorine as a disinfectant.

7.4 TEMPERATURE

Water possesses many important thermal qualities. For instance, water has a high specific heat. Water is not subject to rapid temperature fluctuations, because it can absorb or lose large amounts of heat with relatively small changes in temperature. Water temperature changes gradually in response to seasonal changes. Small water bodies will be influenced by air temperature more quickly than larger water bodies.

The ideal water supply has, at all times, an almost constant temperature or one with minimum variation. Real world conditions, however, do not always provide this, especially in surface water supplies. Thermal pollution is often cited as a cause of wild variations in surface water supplies (see Chapter 10). Many thermal pollution problems are a result of anthropogenic (man-made) activities. However, some water quality problems occur because of natural temperature fluctuations.

Whatever the cause of temperature fluctuations, to live and reproduce fish and other aquatic organisms require certain conditions of temperature. For example, the optimum temperature for trout (a cold water fish) is 15°C. Carp require a temperature of about 32°C—more than twice the preferred temperature for trout.

The problem with heat or temperature in surface waters (besides the health of the fish population) for the drinking water practitioner is that it affects the solubility of oxygen in water, the rate of bacterial activity, and the rate at which gases are transferred to and from the water.

Temperature is not normally used to evaluate water, other than the fact that most people prefer cold drinking water; temperature has little direct significance in public water supplies. Note that temperature is one of the most important parameters in natural surface water systems. Surface waters are subject to great temperature variations.

Water temperature also affects how efficiently certain water treatment processes operate. For example, temperature affects the rate at which chemicals dissolve and react. Cold water requires more chemicals for efficient coagulation and flocculation to take place. When water temperature is high, chlorine demand may also rise because of the increased reactivity, and because warmer temperatures often cause an increased level of algae and other organic matter in raw water.

Note that temperature values are not normally standardized by public health criteria because of the insignificant health effects. However, temperature does have an influence on the treatment of water supplies; on the aquatic life of water reservoirs (biochemical reactions may double the reaction rate for a 10°C increase in temperature); on the taste of drinking water; on the oxygen dissolved; on the activity of organisms producing bad taste and odor; on solubility of solids in water; and on the rate of corrosion of the distribution system (De Zuane, 1997).

Note: When sampling, temperature readings must be done immediately because of changes caused by air temperature and manipulation of the sample.

7.5 TURBIDITY

Turbidity is a unit of measurement quantifying the degree to which light traveling through a water column is scattered by the suspended organic (including algae) and inorganic particles. The scattering of light increases with a greater suspended load. Turbidity is commonly measured in *Nephelometric Turbidity Units* (NTU), but may also be measured in *Jackson Turbidity Units* (JTU).

Note: In the Nephelometric Method, observers compare the light scattered by the sample and the light scattered by a reference solution.

- *detection limits:* should be able to detect turbidity differences of 0.02 NTU with a range of 0 to 40 NTU
- *interferences:* rapidly settling coarse debris, dirty glassware, presence of air bubbles, and surface vibrations

The velocity of the water resource largely determines the composition of the suspended load. Suspended loads are carried in both the gentle currents of lentic (lake) waters and the fast currents of lotic (flowing) waters. Even in flowing water, the suspended load usually consists of grains of less than 0.5 mm in diameter. Suspended loads in lentic waters usually consist of the smallest sediment fraction—silt and clay.

Turbidity plays an important role in drinking water quality, for one of the first things noticed about water is its clarity. Turbidity also may be composed of organic and/or inorganic constituents, and these organic particulates may harbor microorganisms. Thus, turbid conditions may increase the possibility for waterborne disease (see Chapter 6). Turbidity may be classified as both a physical parameter, because it causes aesthetic and psychological objections by the consumer, and as a microbiological parameter, because it may harbor pathogens and impede the effectiveness of disinfection.

Note: Inorganic constituents have no notable health effects.

In surface water supplies, most turbidity results from the erosion of very small colloidal material, including rock fragments, silt, clay, and metal oxides from the soil. Microorganisms and vegetable material may also contribute to turbidity. Wastewaters from industry and households usually contain a wide variety of turbidity-producing material. Detergents, soaps, and various emulsifying agents contribute to turbidity.

Turbidity measurements are normally made on "clean" waters as opposed to wastewaters. In water treatment, turbidity is useful in defining drinking water quality and is relatively easy to measure. Given that the total coliform test is a very reliable routine test of drinking water quality, but not an actual determination of pathogens in water, its use in combination with a turbidity reading and their joint evaluation provide an extra safety factor to the judgment of water quality changes both at the source or during distribution system sampling (De Zuane, 1997).

Note: Formerly, drinking water practitioners in the preliminary evaluation of raw water were of the opinion that when turbidity at the source of supply was under 10 units, disinfection only was required—with BOD at less than 1.0, coliform under 50 MPN/100 mL monthly average, and with acceptable chemical parameters. When turbidity at the source was over 40 units, conventional treatment was considered necessary.

The 1996 Safe Drinking Water Act Amendments and Interim Enhanced Surface Water Treatment (IESWT) rules (a treatment optimization rule), which apply to large (those serving more than 10,000 people) public water systems that use surface water or groundwater directly influenced by surface water regulate turbidity. The rule requires continuous turbidity monitoring of individual filters, and tightens allowable turbidity limits for combined filter effluent, cutting the maximum from 5 NTU to 1 NTU and the average monthly limit from 0.5 NTU (for conventional or direct filtration) to 0.3 NTU in at least 95% of the daily samples for any two consecutive months.

7.6 SOLIDS

All water contaminants other than gases contribute to the solids content. *Solids* can be dispersed in water in both suspended and dissolved forms. Some dissolved solids may be perceived by the physical senses, but fall more appropriately under the category of chemical parameters (see Chapter 8).

Classified by size and state, by chemical characteristics, and by size distribution, solids in drinking water may consist of inorganic particles (salts) with small concentrations of inorganic matter, or of immiscible liquids. Contributory ions are mainly carbonate, bicarbonate, chloride, sulfate, nitrate, potassium, sodium, magnesium, and calcium. Organic material such as plant fibers and biological solids (bacteria, etc.) are also common constituents of surface waters. Inorganics include clay, silt, and other soil constituents common in surface waters. These materials are often natural contaminants resulting from the erosive action of water flowing over surfaces. The filtering properties of soil generally mean that suspended solids are seldom a constituent of groundwater.

Other suspended material may result from human use of the water. For example, domestic wastewater usually contains large quantities of suspended solids that are mostly organic in nature. Industrial use of water may result in a wide variety of organic or inorganic suspended impurities. Immiscible liquids such as oils and greases are often constituents of wastewater.

The solids parameter is used to evaluate and measure all suspended and dissolved matters in water. Solids are classified (in spite of their chemical composition) among the physical parameters of water quality.

In water, suspended material is objectionable because it provides adsorption sites for biological and chemical agents. These adsorption sites provide a protective barrier for attached microorganisms against the chemical action of chlorine disinfectants. Suspended solids in water may be degraded biologically, resulting in objectionable by-products. These factors make the removal of these solids of great concern in the production of clean, safe drinking water and wastewater effluent.

In water treatment, the most effective means of removing solids from water is by filtration. However, some solids, (including colloids and other dissolved solids) cannot be removed by filtration.

Several different tests may be performed on raw and treated waters in relation to solids: dissolved solids, settleable solids, suspended solids, total solids, volatile solids, and conductance.

- *Total dissolved solids (TDS):* TDS in water sample are limited to the solids in solution. TDS recommended upper limit has been 500 mg/L.
- *Settleable solids:* solids in suspension that can be expected to settle by gravity only in a quiescent status like an oversized settling tank. Period of time must be defined. Commonly used in analysis of sewage, this test may indicate data useful to evaluate the sedimentation process, but only when handling very high turbidity.
- *Suspended solids (SS or TSS):* solids not dissolved; also called suspended matter. Little or no significance in water for domestic consumption where turbidity provides a proportional if not equivalent value but with easier determination.
- *Total solids:* all solids contained in the water sample, determined by evaporation and drying.
- *Volatile solids:* solids made up of organic chemicals.
- *Conductance (specific conductance):* a measure of the electric current in the water sample carried by the ionized substances; therefore, dissolved solids are basically related to this measure, which is also influenced by the good conductivity of inorganic acids, bases, and salts, and the poor conductivity of organic compounds.

The methods prescribed by *Standard Methods* (APHA-AWWA-WEF, 1995) for determination of solids are:

- total solids dried at 103°C–105°C
- total dissolved solids dried at 180°C
- total suspended solids dried at 103°C–105°C
- fixed and volatile solids ignited at 550°C
- settleable solids (Imhoff cone—volumetric—gravimetric)
- total, fixed, and volatile solids in solid or semisolid samples

7.7 pH

Raw water examined for potential use as drinking water has an expected pH value between 4 and 9, but, more than likely, encountered values will be between 5.5 and 8.6.

What does this mean?

pH is defined as the negative log-base 10 of the hydrogen ion concentration:

$$pH = -\log 10[H^+] \tag{7.2}$$

Since the pH is a log-base-10 scale, the pH changes one unit for every power of ten change in $[H^+]$. For example, a water with a pH of 3 has 100 times the amount of [H^+] found in a pH 5 water. Remember that because pH = $-\log 10$ [H^+], the pH will decrease as the [H^+] increases.

Water's pH is controlled by the equilibrium achieved by dissolved compounds in the system. In natural waters, the pH is primarily a function of the carbonate system, which is composed of carbon dioxide, carbonic acid, bicarbonate, and carbonate. Acid inputs to a water system may substantially alter the pH. The main sources of acid include acid mine drainage and atmospheric acid deposition.

Low pH water may corrode distribution pipes in potable water plants. The pipes may be costly to replace and the corrosion may release metal ions such as copper, lead, zinc, and cadmium into the treated drinking water. Ingestion of heavy metals may pose substantial health risks to humans. The minimum and maximum allowable pH range for potability is 6.5–8.5 (Safe Drinking Water Act).

Note: The role of pH in water is also associated with corrosivity, hardness, acidity, chlorination, coagulation, carbon dioxide stability, and alkalinity.

7.8 ALKALINITY

Alkalinity is a measure of water's ability to absorb hydrogen ions without significant pH change. Simply stated, alkalinity is a measure of the buffering capacity of water, and is thus a measure of the ability of water to neutralize acids. The major chemical constituents of alkalinity in natural water supplies are bicarbonate, carbonate, and hydroxyl ions. These compounds are mostly the carbonates and bicarbonates of sodium, potassium, magnesium, and calcium. These constituents originate from carbon dioxide (from the atmosphere and as a by-product of microbial decomposition of organic material) and from their mineral origin (primarily from chemical compounds dissolved from rocks and soil).

Highly alkaline water is unpalatable; however, this condition has little known significance on human health. The principal problem with alkaline water is the reactions that occur between alkalinity and certain substances in the water. The resultant precipitate can foul water system appurtenances. Alkalinity levels also affect the efficiency of certain water treatment processes, especially the coagulation process.

Note: Total alkalinity is determined by titration with sulfuric acid or other strong acids of known strength to the end point of indicators [see *Standard Methods* (APHA-AWWA-WEF, 1995), Appendix A-VII, "General References"], with the result expressed in mg/L of calcium carbonate equivalent to the determined alkalinity.

7.9 HARDNESS

Water *hardness* is commonly defined as the sum of the polyvalent cations dissolved in the water. The most common cations are calcium and magnesium, although iron, strontium, and manganese may contribute. Hardness is usually reported as an equivalent quantity of calcium carbonate. Generally, waters are classified according to degree of hardness as follows:

Concentration Calcium Carbonate (mg/L)	Classification
<75	soft water
75–150	moderately hard
150–300	hard
>300	very hard

Hardness is primarily a function of the geology of the area with which the surface water is associated. Waters underlain by limestone are prone to hard water because rainfall (naturally acidic because it contains carbon dioxide gas) continually dissolves the rock and carries the dissolved cations into the water system.

Standard Methods (APHA-AWWA-WEF, 1995) recommends measuring hardness by using the calculation in Equation (7.3):

$$Hardness\,(mg\,/\,L) = 2497\,(Ca, mg\,/\,L) + 4.118\,(Mg, mg\,/\,L) \quad (7.3)$$

Hardness can also be measured by using the *EDTA titration method* [see *Standard Methods* (APHA-AWWA-WEF, 1995)].

7.10 WATER SOLUBILITY

Solubility is a term often used in connection with water treatment, in laboratory analyses, in chemical/physical studies of water, and in related technical publications, even though solubility is not a general, physical, or chemical parameter.

To understand solubility, you must also understand *water solution*, a homogeneous liquid made of the *solvent* (the substance that dissolves another substance) and the *solute* (the substance that dissolves in the solvent).

Solubility is defined as the mass of substance contained in the solution that is in equilibrium with an excess of the substance.

7.11 SUMMARY

Water's biological and physical characteristics are two thirds of the parameters critical to understanding the elements drinking water technologists may encounter. The final third that makes up the whole is presented in Chapter 8.

7.12 REFERENCES

APHA-AWWA-WEF. *Standard Methods for the Examination of Water and Wastewater.* 19th ed. Washington, D.C.: American Public Health Association, 1995.

Clary, D., "What Makes Water Wet." *Geraghty and Miller Water Newsletter*, 39:4, 1997.

De Zuane, J., *Handbook of Drinking Water Quality.* 2nd ed. New York: John Wiley and Sons, Inc., 1997.

Lawrence, D. H. "The Third Thing." *Pansies,* 1929.

Spellman, F. R., *The Science of Water: Concepts and Applications.* Lancaster, PA: Technomic Publishing Company, Inc., 1998.

USEPA. National Secondary Drinking Water Regulations—Part 143. *Federal Register,* Vol. 54, No. 97, May 22, 1989.

CHAPTER 8

Drinking Water Parameters: Chemical

Water, in any of its forms, also . . . [has] scant respect for the laws of chemistry.

Most materials act either as acids or bases, settling on either side of a natural reactive divide. Not water. It is one of the few substances that can behave both as an acid and as a base, so that under certain conditions it is capable of reacting chemically with itself. Or with anything else.

Molecules of water are off balance and hard to satisfy. They reach out to interfere with every other molecule they meet, pushing its atoms apart, surrounding them, and putting them into solution. Water is the ultimate solvent, wetting everything, setting other elements free from the rocks, making them available for life. Nothing is safe. There isn't a container strong enough to hold it. (L. Watson, The Water Planet, 1988, p. 37)

8.1 INTRODUCTION

WATER chemical parameters are categorized into two basic groups: inorganic and organic chemicals. Both inorganic and organic chemicals enter water from natural causes or pollution.

Note: The solvent capabilities of water are directly related to its chemical parameters.

In this chapter, we do not look at each organic/inorganic chemical individually. Instead, we look at general chemical parameter categories such as dissolved oxygen organics (BOD and COD), (DO), synthetic organic chemicals (SOCs), volatile organic chemicals (VOCs), total dissolved solids (TDS), fluorides, metals, and nutrients—the major chemical parameters of concern.

8.2 ORGANICS

Natural organics contain carbon and consist of biodegradable organic

153

matter such as wastes from biological material processing, human sewage, and animal feces. Microbes aerobically break down the complex organic molecules into simpler, more stable end products. Microbial degradation yields end products such as carbon dioxide, water, phosphate, and nitrate. Organic particles in water may harbor harmful bacteria and pathogens. Infection by microorganisms may occur if the water is used for primary contact or as a raw drinking water source. Treated drinking water will not present the same health risks. In a potable drinking water plant, all organics should be removed in the water before disinfection (see Chapter 11).

Organic chemicals also contain carbon; they are substances that come directly from, or are manufactured from, plant or animal matter. Plastics provide a good example of organic chemicals that are made from petroleum, which originally came from plant and animal matter. Some organic chemicals (like those discussed above) released by decaying vegetation, occur naturally and by themselves tend not to pose health problems when they get in our drinking water. However, more serious problems are caused by the more than 100,000 different manufactured or synthetic organic chemicals in commercial use today. They include paints, herbicides, synthetic fertilizers, pesticides, fuels, plastics, dyes, preservatives, flavorings, and pharmaceuticals, to name a few.

Many organic materials are soluble in water, are toxic, and are found in public water supplies. According to Tchobanoglous and Schroeder (1987), the presence of organic matter in water is troublesome. Organic matter causes: "(1) color formation, (2) taste and odor problems, (3) oxygen depletion in streams, (4) interference with water treatment process, and (5) the formation of halogenated compounds when chlorine is added to disinfect water" (p. 94).

Remember, organics in natural water systems may come from natural sources or may result from human activities. Generally, the principal source of organic matter in water is from natural sources including decaying leaves, weeds, and trees; the amount of these materials present in natural waters is usually low. Anthropogenic (man-made) sources of organic substances come from pesticides and other synthetic organic compounds.

Again, many organic compounds are soluble in water, and surface waters are more prone to contamination by natural organic compounds than are groundwaters. In water, dissolved organics are usually divided into two categories: *biodegradable* and *nonbiodegradable*.

Biodegradable (able to break down) material consists of organics that can be used for food (nutrients) by naturally occurring microorganisms within a reasonable length of time. Alcohols, acids, starches, fats, proteins, esters, and aldehydes are the main constituents of biodegradable materials. They may result from domestic or industrial wastewater discharges, or they may be end products of the initial microbial decomposition of plant or animal tissue. Biodegradable organics in surface waters cause problems mainly associated with the effects that result from the action of microorganisms. As the microbes metabolize organic material, they consume oxygen.

When this process occurs in water, the oxygen consumed is dissolved oxygen (DO). If the oxygen is not continually replaced in the water by artificial means, the DO level will decrease as the organics are decomposed by the microbes. This need for oxygen is called the *biochemical oxygen demand* (BOD): the amount of dissolved oxygen demanded by bacteria to break down the organic materials during the stabilization action of the decomposable organic matter under aerobic conditions over a five-day incubation period at 20°C (68°F). This bioassay test measures the oxygen consumed by living organisms using the organic matter contained in the sample and dissolved oxygen in the liquid. The organics are broken down into simpler compounds and the microbes use the energy released for growth and reproduction. A BOD test is not required for monitoring drinking water.

Note: The more organic material in the water, the higher the BOD exerted by the microbes will be.

Note also that some biodegradable organics can cause color, taste, and odor problems.

Nonbiodegradable organics are resistant to biological degradation. The constituents of woody plants are a good example. These constituents, including tannin and lignic acids, phenols, and cellulose are found in natural water systems, and are considered refractory (resistant to biodegradation). Some polysaccharides with exceptionally strong bonds, and benzene (for example, associated with the refining of petroleum) with its ringed structure are essentially nonbiodegradable.

Certain nonbiodegradable chemicals can react with oxygen dissolved in water. The *Chemical Oxygen Demand* (COD) is a more complete and accurate measurement of the total depletion of dissolved oxygen in water. *Standard Methods* (1999) defines COD as a test that provides a measure of the oxygen equivalent of that portion of the organic matter in a sample that is susceptible to oxidation by a strong chemical oxidant. The procedure is detailed in *Standard Methods* (1999).

Note: COD is not normally used for monitoring water supplies, but is often used for evaluating contaminated raw water.

8.3 SYNTHETIC ORGANIC CHEMICALS (SOCs)

Synthetic organic chemicals (SOCs) are man-made, and because they don't occur naturally in the environment, they are often toxic to humans. More than 50,000 SOCs are in commercial production, including common pesticides, carbon tetrachloride, chloride, dioxin, xylene, phenols, aldicarb, and thousands of others. Unfortunately, even though they are so prevalent,

little data has been collected on these toxic substances. Determining defini-
tively just how dangerous many of the SOCs are is rather difficult.

8.4 VOLATILE ORGANIC CHEMICALS (VOCs)

Volatile organic chemicals (VOCs) are a type of organic chemical that is
particularly dangerous. VOCs are absorbed through the skin during contact
with water—as in the shower or bath. Hot water exposure allows these
chemicals to evaporate rapidly, and they are harmful if inhaled. VOCs can
be in any tap water, regardless of what part of the country one lives in and
the water supply source.

8.5 TOTAL DISSOLVED SOLIDS (TDS)

Earlier we pointed out that solids in water occur either in solution or in
suspension, and are distinguished by passing the water sample through a
glass-fiber filter. By definition, the *suspended solids* are retained on top of
the filter, and the *dissolved solids* pass through the filter with the water. When
the filtered portion of the water sample is placed in a small dish and then
evaporated, the solids in the water remain as residue in the evaporating dish.
This material is called *total dissolved solids*, or TDS.

Dissolved solids may be organic or inorganic. Water may come into
contact with these substances within the soil, on surfaces, and in the atmosphere. The organic dissolved constituents of water are from the decay
products of vegetation, from organic chemicals, and from organic gases.
Removing these dissolved minerals, gases, and organic constituents is
desirable, because they may cause physiological effects and produce aesthetically displeasing color, taste, and odors.

Note: In water distribution systems, a high TDS means high conductivity
with consequent higher ionization in corrosion control. However, high TDS
also means more likelihood of a protective coating, a positive factor in
corrosion control.

8.6 FLUORIDES

According to Phyllis J. Mullenix (1997), water fluoridation is not the safe
public health measure we have been led to believe. Concerns about uncontrolled dosage, accumulation in the body over time, and effects beyond the
teeth (brain as well as bones) have not been resolved for fluoride. The health
of citizens necessitates that all the facts be considered, not just those that are
politically expedient.

Most medical authorities would take issue with Mullenix's view on the
efficacy of fluoride in reducing tooth decay. Most authorities seem to hold
that a moderate amount of fluoride ions (F⁻) in drinking water contributes to

good dental health. Fluoride is seldom found in appreciable quantities of surface waters and appears in groundwater in only a few geographical regions, though it is sometimes found in a few types of igneous or sedimentary rocks. Fluoride is toxic to humans in large quantities (the key words are "large quantities" or in Mullenix's view "uncontrolled dosages") and also toxic to some animals.

Fluoride used in small concentrations (about 1.0 mg/L in drinking water) can be beneficial. Experience has shown that drinking water containing a proper amount of fluoride can reduce tooth decay by 65% in children between the ages of 12 and 15. However, when the concentration of fluorides in untreated natural water supplies is excessive, either alternative water supplies must be used, or treatment to reduce the fluoride concentration must be applied, because excessive amounts of fluoride cause mottled or discolored teeth, a condition called *dental fluorosis*.

8.7 HEAVY METALS

Heavy metals are elements with atomic weights between 63.5 and 200.5, and a specific gravity greater than 4.0. Living organisms require trace amounts of some heavy metals, including cobalt, copper, iron, manganese, molybdenum, vanadium, strontium, and zinc. Excessive levels of essential metals, however, can be detrimental to the organism. Nonessential heavy metals of particular concern to surface water systems are cadmium, chromium, mercury, lead, arsenic, and antimony.

Heavy metals in water are classified as either nontoxic or toxic. Only those metals that are harmful in relatively small amounts are labeled toxic; other metals fall into the nontoxic group. In natural waters (other than in groundwaters), sources of metals include dissolution from natural deposits and discharges of domestic, agricultural, or industrial wastes.

All heavy metals exist in surface waters in colloidal, particulate, and dissolved phases, although dissolved concentrations are generally low. The colloidal and particulate metal may be found in (1) hydroxides, oxides, silicates, or sulfides; or (2) adsorbed to clay, silica, or organic matter. The soluble forms are generally ions or deionized organometallic chelates or complexes. The solubility of trace metals in surface waters is predominately controlled by water pH, the type and concentration of liquids on which the metal could adsorb, and the oxidation state of the mineral components and the redox environment of the system.

The behavior of metals in natural waters is a function of the substrate sediment composition, the suspended sediment composition, and the water chemistry. Sediment composed of fine sand and silt will generally have higher levels of adsorbed metal than will quartz, feldspar, and detrital carbonate-rich sediment.

The water chemistry of the system controls the rate of adsorption and desorption of metals to and from sediment. Adsorption removes the metal from the water column and stores the metal in the substrate. Desorption returns the metal to the water column, where recirculation and bioassimilation may take place. Metals may be desorbed from the sediment if the water experiences increases in salinity, decreases in redox potential, or decreases in pH.

Although heavy metals such as iron (Fe) and manganese (Mn) do not cause health problems, they do impart a noticeable bitter taste to drinking water, even at very low concentrations. These metals usually occur in groundwater in solution, and these and others may cause brown or black stains on laundry and on plumbing fixtures.

8.8 NUTRIENTS

Elements in water (such as carbon, nitrogen, phosphorous, sulfur, calcium, iron, potassium, manganese, cobalt, and boron—all essential to the growth and reproduction of plants and animals) are called *nutrients* (or biostimulants). The two nutrients that concern us in this text are nitrogen and phosphorous.

Nitrogen (N_2), an extremely stable gas, is the primary component of the Earth's atmosphere (78%). The nitrogen cycle is composed of four processes. Three of the processes—fixation, ammonification, and nitrification—convert gaseous nitrogen into usable chemical forms. The fourth process—denitrification—converts fixed nitrogen back to the unusable gaseous nitrogen state.

Nitrogen occurs in many forms in the environment and takes part in many biochemical reactions. Major sources of nitrogen include runoff from animal feedlots, fertilizer runoff from agricultural fields, municipal wastewater discharges, and certain bacteria and blue-green algae that obtain nitrogen directly from the atmosphere. Certain forms of acid rain can also contribute nitrogen to surface waters.

Nitrogen in water is commonly found in the form of *nitrate* (NO_3), which indicates that the water may be contaminated with sewage. Nitrates can also enter the groundwater from chemical fertilizers used in agricultural areas. Excessive nitrate concentrations in drinking water pose an immediate health threat to infants, both human and animal, and can cause death. The bacteria commonly found in the intestinal tract of infants can convert nitrate to high toxic nitrites (NO_2). Nitrite can replace oxygen in the bloodstream and results in oxygen starvation that causes a bluish discoloration of the infant ("blue baby" syndrome).

Note: Lakes and reservoirs usually have less than 2 mg/L of nitrate measured as nitrogen. Higher nitrate levels are found in groundwater ranging up to 20 mg/L, but much higher values are detected in shallow aquifers polluted by sewage and/or excessive use of fertilizers.

Phosphorous (P) is an essential nutrient that contributes to the growth of algae and the eutrophication of lakes, though its presence in drinking water has little effect on health. In aquatic environments, phosphorous is found in the form of phosphate and is a limiting nutrient. If all phosphorous is used, plant growth ceases, no matter the amount of nitrogen available. Many bodies of freshwater currently experience influxes of nitrogen and phosphorous from outside sources. The increasing concentration of available phosphorous allows plants to assimilate more nitrogen before the phosphorous is depleted. If sufficient phosphorous is available, high concentrations of nitrates will lead to phytoplankton (algae) and macrophyte (aquatic plant) production.

Major sources of phosphorous include phosphates in detergents, fertilizer and feedlot runoff, and municipal wastewater discharges. The USEPA 1976 *Water Quality Standards—Criteria Summaries for Phosphorous*—recommended a phosphorous criterion of 0.10 µg/L (elemental) phosphorus for marine and estuarine waters, but no freshwater criterion.

8.9 SUMMARY

The biological, physical, and chemical condition of our water is of enormous concern to us all, because we must live in such intimate contact with water. When these parameters shift and change, the changes affect us, often in ways science cannot yet define. Water pollution is an external element that can and does significantly affect our water. Water pollution, however, doesn't always go straight from source to water. Controlling what goes into our water is difficult, because the hydrologic cycle carries water (and whatever it picks up along the way) through all of our environment's media, affecting the biological, physical, and chemical condition of the water we must drink to live. We discuss water pollution further in Chapter 9.

8.10 REFERENCES

Mullenix, P. J., in a letter sent to Operations and Environmental Committee, City of Calgary, Canada, 1997.

Standard Methods for Examination of Water and Wastewater, 20th Edition, eds., A. E. Greenberg et al. Washington, D.C.: American Public Health Assn., 1999.

Tchobanoglous, G. and Schroeder, E. D., *Water Quality.* Reading, Massachusetts: Addison-Wesley Publishing Company, 1987.

USEPA, *Water Quality Standards—Criteria Summaries for Phosphorous.* Washington, D. C.: USEPA, p. 17, 1976.

Watson, L., *The Water Planet: A Celebration of the Wonders of Water.* New York: Crown Publishers, Inc., 1988.

CHAPTER 9

Water Pollution

This country's waterways have been transformed by omission. Without bea-
vers, water makes its way too quickly to the sea; without prairie dogs, water
runs over the surface instead of sinking into the aquifer; without bison, there
are no groundwater-recharge ponds in the grasslands and the riparian zone
is trampled; without alligators, the edge between the water and land is
simplified. Without forests, the water runs unfiltered to the waterways, and
there is less deadwood in the channel, reducing stream productivity. Without
floodplains and meanders, the water moves more swiftly, and silt carried in
the water is more likely to be swept to sea.

The beaver, the prairie dog, the bison, and the alligator have been scarce for
so long that we have forgotten how plentiful they once were. Beaver popula-
tions are controlled, because they flood fields and forests, while wetlands
acreage decrease annually. Prairie dogs are poisoned, because they compete
with cattle for grass, while the grasslands grow more barren year by year.
Buffalo are generally seen as photogenic anachronisms, and alligators are
too reptilian to be very welcome. But all of these animals once shaped the
land in ways that improve water quality. (Alice Outwater, Water: A Natural
History, *pp. 175–176, 1996)*

9.1 INTRODUCTION

IS drinking water contamination really a problem—a serious problem? The
answer to the first part of the question depends upon where your water
comes from. As to the second part of the question, we refer you to a book
(or the film based upon the book) that concerns a case of toxic contamination,
A Civil Action, written by Jonathan Harr. The book and film portray the legal
repercussions connected with polluted water supplies in Woburn, Massachusetts. Two wells became polluted with industrial solvents, in all apparent
likelihood causing 24 of the town's children, who lived in neighborhoods
supplied by those wells, to contract leukemia and die.

Many who have read the book or have seen the movie may mistakenly get the notion that Woburn, a toxic "hot spot," is a rare occurrence. Nothing could be further from the truth. Toxic "hot spots" abound. Most striking are areas of cancer clusters. A short list includes:

- Woburn, where about two dozen children were stricken with leukemia over 12 years, a rate several times the national average for a community of its size.
- Storrs, Connecticut, where wells polluted by a landfill are suspected of sickening and killing residents in nearby homes.
- Bellingham, Washington, where pesticide-contaminated drinking water is thought to be linked to a sixfold increase in childhood cancers.

As Schlichtmann[11] points out, these are only a few examples of an underlying pathology that threatens many other communities. Meanwhile, cancer is now the primary cause of childhood death from disease.

Drinking water contamination is a problem—a very serious problem. In this chapter, we discuss a wide range of water contaminants, the contaminant sources, and their impact on drinking water supplies from both surface water and groundwater sources.

9.2 SOURCES OF CONTAMINANTS

If we were to list all the sources of contaminants and the contaminants themselves (the ones that can and do foul our water supply systems), along with a brief description of each contaminant, we could easily fill a book. To give you some idea of the magnitude of the problem we present a condensed list of selected sources and contaminants.

Note: Keep in mind that when we specify "water pollutants" we are in most cases speaking about pollutants that somehow get into the water (by whatever means) from the interactions of the other two environmental mediums: air and soil. Probably the best example of this is the acid rain phenomenon; pollutants originally emitted only into the atmosphere land on earth and affect both soil and water. Consider that 69% of the anthropogenic lead and 73% of the mercury in Lake Superior reach it by atmospheric deposition (Hill, 1997).

(1) Subsurface percolation—hydrocarbons, metals, nitrates, phosphates, microorganisms, and cleaning agents [trichloroethylene (TCE)]
(2) Injection wells—hydrocarbons, metals, nonmetals, organics, organic and inorganic acids, microorganisms, and radionuclides

---

[11]This account, written by Jan Schlichtmann, appeared in *USA Today*, February 4, 1999.

(3) Land application—nitrogen, phosphorous, heavy metals, hydrocarbons, microorganisms, and radionuclides

(4) Landfills—organics, inorganics, microorganisms, and radionuclides

(5) Open dumps—organics, inorganics, and microorganisms

(6) Residential (local) disposal—organic chemicals, metals, nonmetal inorganics, inorganic acids, and microorganisms

(7) Surface impoundments—organic chemicals, metals, nonmetal inorganics, inorganic acids, microorganisms, and radionuclides

(8) Waste tailings—arsenic, sulfuric acid, copper, selenium, molybdenum, uranium, thorium, radium, lead, manganese, vanadium

(9) Waste piles—arsenic, sulfuric acid, copper, selenium, molybdenum, uranium, thorium, radium, lead, manganese, vanadium

(10) Materials stockpiles—coal pile: aluminum, iron, calcium, manganese, sulfur, and traces of arsenic, cadmium, mercury, lead, zinc, uranium, and copper; other materials piles: metals/nonmetals and microorganisms

(11) Graveyards—metals, nonmetals, and microorganisms

(12) Animal burial—contamination is site-specific, depending on disposal practices, surface and subsurface, hydrology, proximity of the site to water sources, type and amount of disposed material, and cause of death

(13) Above ground storage tanks—organics, metal/nonmetal inorganics, inorganic acids, microorganisms, and radionuclides

(14) Underground storage tanks—organics, metal, inorganic acids, microorganisms, and radionuclides

(15) Containers—organics, metal/nonmetal inorganics, inorganic acids, microorganisms, and radionuclides

(16) Open burning and detonating sites—inorganics, including heavy metals; organics, including TNT

(17) Radioactive disposal sites—radioactive cesium, plutonium, strontium, cobalt, radium, thorium, and uranium

(18) Pipelines—organics, metals, inorganic acids, and microorganisms

(19) Material transport and transfer operations—organics, metals, inorganic acids, microorganisms, and radionuclides

(20) Irrigation practices—fertilizers, pesticides, naturally occurring contamination and sediments

(21) Pesticide applications—1200–1400 active ingredients; contamination already detected: alachlor, aldicarb, atrazine, bromacil, carbofuran, cyanazine, DBCP, DCPA, 1,2-dichloropropane, dyfonate, EDB, metolachlor, metribyzen, oxalyl, siazine, and 1,2,3-trichloropropane (The extent of groundwater contamination cannot be determined with current data.)

(22) Animal feeding operations—nitrogen, bacteria, viruses, and phosphates

(23) De-icing salts applications—chromate, phosphate, ferric ferrocyanide, chlorine

(24) Urban runoff—suspended solids and toxic substances, especially heavy metals and hydrocarbons, bacteria, nutrients, and petroleum residues

(25) Percolation of atmospheric pollutants—sulfur and nitrogen compounds, asbestos, and heavy metals

(26) Mining and mine drainage—
  - coal: acids, toxic inorganics (heavy metals), and nutrients phosphate: radium, uranium, and fluorides
  - metallic ores: sulfuric acid, lead, cadmium, arsenic, sulfur, cyan

(27) Production wells—
  - oil: 1.2 million abandoned production wells
  - irrigation: farms
  - all: potential to contaminate—installation, operation, and plugging techniques

(28) Construction excavation—pesticides, diesel fuel, oil, salt, and variety of others

Note: Before we discuss specific water pollutants, we must examine several terms important to the understanding of water pollution. One of these is *point source*. The USEPA defines a *point source* as "any single identifiable source of pollution from which pollutants are discharged, e.g., a pipe, ditch, ship, or factory smokestack." For example, the outlet pipes of an industrial facility or a municipal wastewater treatment plant are point sources. In contrast, *nonpoint sources* are widely dispersed sources and are a major cause of stream pollution. An example of a nonpoint source of pollution is rainwater carrying topsoil and chemical contaminants into a river or stream. Some of the major sources of nonpoint pollution include water runoff from farming, urban areas, forestry, and construction activities. The word *runoff* signals a nonpoint source that originated on land. Runoff may carry a variety of toxic substances and nutrients, as well as bacteria and viruses with it. Nonpoint sources now comprise the largest source of water pollution, contributing approximately 65% of the contamination in quality-impaired streams and lakes.

9.3 RADIONUCLIDES

When radioactive elements decay, they emit alpha, beta, or gamma radiations caused by transformation of the nuclei to lower energy states. In drinking water, radioactivity can be from natural or artificial radionuclides (the radioactive metals and minerals that cause contamination). These radioactive substances in water are of two types: radioactive minerals and radioactive gas. The USEPA reports that some 50 million Americans face increased cancer risk because of radioactive contamination of their drinking water.

Because of their occurrence in drinking water and their effects on human health, the natural radionuclides of chief concern are radium-226, radium-228, radon-222, and uranium. The source of some of these naturally occurring radioactive minerals is typically associated with certain regions of the country where mining is active or was active in the past. Mining activities expose rock strata, most of which contains some amount of radioactive ore. Uranium mining, for example, produces runoff. Another source of natural radioactive contamination occurs when underground streams flow through various rockbed and geologic formations containing radioactive materials. Other natural occurring sources where radioactive minerals may enter water supplies are smelters and coal-fired electrical generating plants.

Sources of man-made radioactive minerals in water are nuclear power plants, nuclear weapons facilities, radioactive materials disposal sites, and mooring sites for nuclear-powered ships. Hospitals also contribute radioactive pollution when they dump low-level radioactive wastes into sewers. Some of these radioactive wastes eventually find their way into water supply systems.

While radioactive minerals such as uranium and radium in water may present a health hazard in these particular areas, a far more dangerous threat exists in the form of radon. *Radon* is a colorless, odorless gas created by (a by-product of) the natural decay of minerals in the soil. Normally present in all water in minute amounts, radon is especially concentrated in water that has passed through rock strata of granite, uranium, or shale. Radon enters homes from the soil beneath the house, through cracks in the foundation, through crawl spaces and unfinished basements, and in tainted water, and is considered the second leading cause of lung cancer in the United States (about 20,000 cases each year), following cigarette smoking. Contrary to popular belief, radon is not a threat from surface water (lake, river, or above-ground reservoir), because radon dissipates rapidly when water is exposed to air. Even if the water source is groundwater, radon is still not a threat if the water is exposed to air (aerated) or if it is processed through an open tank during treatment. Studies show that, where high concentrations of radon occur within the air in a house, most of the radon comes through the foundation and from the water. However, radon in the tap water, showers, baths, and cooking (with hot water) will cause high concentrations of radon in the air.

Note, however, that radon is a threat from groundwater taken directly from an underground source—either a private well or from a public water supply whose treatment of the water does not include exposure to air. Because radon in water evaporates quickly into air, the primary danger is from inhaling it from the air in a house, not from drinking it.

9.4 THE CHEMICAL COCKTAIL

In previous pages, we referred to a glass of water filled to the brim from the household tap. When we hold such a full glass and inspect the contents, a few possibilities might present themselves. The contents might appear cloudy or colored (making us think that the water is not fit to drink). The contents might look fine, but carry the prevalent odor of chlorine. Most often, water drawn from the tap will simply look like water and we will drink it, use it to cook with, or whatever.

The fact is, typically a glass of treated water is a chemical cocktail (Kay, 1996). While water utilities in communities seek to protect the public health by treating raw water with certain chemicals, what they are in essence doing is providing us a drinking water product that is a mixture of various treatment chemicals and their by-products. For example, the water treatment facility typically adds chlorine to disinfect—chlorine can produce contaminants. Another concoction is formed when ammonia is added to disinfect. Alum and polymers are added to the water to settle out various contaminants. The water distribution system and appurtenances also need to be protected to prevent pipe corrosion or soften water, so the water treatment facility adds caustic soda, ferric chloride, and lime, which in turn work to increase the aluminum, sulfates, and salts in the water. Thus, when we hold that glass of water before us, and we perceive what appears to be a full glass of crystal clear, refreshing water, what we really see is a concoction of many chemicals mixed with water, forming that chemical cocktail.

The most common chemical additives used in water treatment are chlorine, fluorides, and flocculants. Because we have already discussed fluorides, we focus our discussion in the following sections on the by-products of chlorine and flocculant additives.

9.4.1 BY-PRODUCTS OF CHLORINE

To lessen the potential impact of that water cocktail, the biggest challenge today is to make sure the old standby—chlorine—won't produce as many new contaminants as it destroys. At the present time, arguing against chlorine and the chlorination process is difficult. Since 1908, chlorine has been used in the United States to kill off microorganisms that spread cholera, typhoid fever, and other waterborne diseases. However, in the 1970s, scientists discovered that while chlorine does not seem to cause cancer in lab animals, it can—in the water treatment process—create a whole list of by-products that do. The by-products of chlorine—organic hydrocarbons called *trihalomethanes* (usually discussed as total trihalomethanes or TTHMs)—present the biggest health concern.

The USEPA classifies three of these trihalomethanes by-products—chloroform, bromoform, and bromodichloromethane—as probable human carcinogens. The fourth, dibromochloromethane, is classified as a possible human carcinogen.

The USEPA set the first trihalomethane limits in 1979. Most water companies met these standards initially, but the standards were tightened after the 1996 SDWA Amendments. The USEPA is continuously studying the need to regulate other cancer-causing contaminants, including haloacetic acids (HAAs) also produced by chlorination.

Most people concerned with protecting public health applaud the USEPA's efforts in regulating water additives and disinfection by-products. However, some of those in the water treatment and supply business express concern. A common concern often heard from water utilities having a tough time balancing the use of chlorine without going over the regulated limits revolves around the necessity of meeting regulatory requirements by lowering chlorine amounts to meet by-products standards, and at the same ensuring that all the pathogenic microorganisms are killed off. Many make the strong argument that while no proven case exists that disinfection by-products cause cancer in humans, many cases—a whole history of cases—show that if we don't chlorinate water, people get sick and sometimes die from waterborne disease.

Because chlorine and chlorination is now prompting regulatory pressure and compliance with new, demanding regulations, many water treatment facilities are looking for options. Choosing an alternative disinfection chemical process is feeding a growing enterprise. One alternative that is currently being given widespread consideration in the U.S. is ozonation, which uses ozone gas to kill microorganisms. Ozonation is Europe's favorite method, and it doesn't produce trihalomethanes. But the USEPA doesn't yet recommend wholesale switchover to ozone to replace chlorine or chlorination systems (sodium hypochlorite or calcium hypochlorite vice elemental chlorine). The USEPA points out that ozone also has problems: it does not produce a residual disinfectant in the water distribution system; it is much more expensive; and in salty water, it can produce another carcinogen, bromate. We discuss disinfection alternatives in greater detail in Chapter 11.

At the present, what drinking water practitioners are doing (in the real world) is attempting to fine-tune water treatment. What it all boils down to is a delicate balancing act. The drinking water professionals don't want to cut back disinfection; if anything, they'd prefer to strengthen it.

How do we bring into parity the microbial risks versus the chemical risks?

How do we reduce them both to an acceptable level? The answer? No one is quite sure how to do this. The problem really revolves around the enigma associated with a "we don't know what we don't know" scenario.

The disinfection by-products problem stems from the fact that most U.S. water systems produce the unwanted by-products when the chlorine reacts to decayed organics: vegetation and other carbon-containing materials in water. Communities that take drinking water from lakes and rivers have a tougher time keeping the chlorine by-products out of the tap than those that use clean groundwater.

In some communities, when a lot of debris is in the reservoir, the water utility switches to alternate sources—wells, for example. In other facilities, chlorine is combined with ammonia in a disinfection method called *chloramination*. This method is not as potent as pure chlorination, but stops the production of unwanted trihalomethanes.

In communities where rains wash leaves, trees, and grasses into the local water source (lake or river), hot summer days trigger algae blooms, upping the organic matter that can produce trihalomethanes. Spring runoff in many communities acerbates the problem. With increased runoff comes agricultural waste, pesticides, and quantities of growth falling into the water that must be dealt with.

Nature's conditions in summer diminish some precursors for trihalomethanes—the bromides in salty water.

The irony is that under such conditions nothing unusual is visible in the drinking water. However, cloudy water from silt (dissolved organics from decayed plants) is enough to create trihalomethanes.

With the advent of the new century, most cities will strain out the organics from their water supplies before chlorinating to prevent the formation of trihalomethanes and haloacetic acids.

In other communities, the move is already on to switch from chlorine to ozone and other disinfectant methods. The National Resources Defense Council states that probably in 15 to 20 years, most U.S. systems will catch up with Europe and use ozone to kill resistant microbes like *Cryptosporidium*. Note that when this method is employed, the finishing touch is usually accomplished by filtering through granulated activated carbon, which increases the cost slightly (estimated at about $100 or more per year per hookup) that customers must pay.

9.4.1.1 Trihalomethane Regulations[12]

To ensure safe drinking water supplies in the U.S., in December 1998, President Clinton announced these rules:

---

[12]The information contained in this section is from USEPA's *Drinking Water Priority Rulemaking: Microbial and Disinfection By-products Rules*, United States Environmental Protection Agency, EPA 815-F-95-0014, 1998.

- Stage 1 Disinfectants and Disinfection By-products Rule: *Federal Register* Notice (HTML)
- Interim Enhanced Surface Water Treatment Rule: *Federal Register* Notice (HTML)

A major challenge for drinking water practitioners is how to balance the risks from microbial pathogens and disinfection by-products. Providing protection from these microbial pathogens while simultaneously ensuring decreasing health risks to the population from disinfection by-products (DBPs) is important. The Safe Drinking Water Act (SDWA) Amendments, signed by the President in August 1996, required the USEPA to develop rules to achieve these goals. The new Stage 1 Disinfectants and Disinfection By-products Rule and Interim Enhanced Surface Water Treatment Rule are the first of a set of rules under the Amendments.

These new rules are a product of six years of collaboration among the water industry, environmental and public health groups, and local, state, and federal government. The schedule for the Microbial-Disinfectants and Disinfection By-products (M-DBP) Rules is:

Schedule of M-DBP Rules

November 1998—Final Rule

Interim Enhanced Surface Water Treatment Rule and Stage 1 Disinfection By-products Rule

August 2000—Final Rule—Filter Backwash Recycling Rule

November 2000—Final Rule—Long Term 1 Enhanced Surface Water Treatment Rule and Groundwater Rule

May 2002—Final Rule—Stage 2 Disinfection By-products Rule and Long Term 2 Enhanced Surface Water Treatment Rule

9.4.1.2 Public Health Concerns

Most Americans drink tap water that meets all existing health standards all the time. These new rules will further strengthen existing drinking water standards and thus increase protection for many water systems.

The USEPA's Science Advisory Board concluded in 1990 that exposure to microbial contaminants such as bacteria, viruses, and protozoa (e.g., *Giardia lamblia* and *Cryptosporidium*) was likely the greatest remaining health risk management challenge for drinking water suppliers. Acute health effects from exposure to microbial pathogens is documented, and associated illness can range from mild to moderate cases lasting only a few days to more severe infections that can last several weeks and may result in death for those with weakened immune systems.

While disinfectants are effective in controlling many microorganisms, they react with natural organic and inorganic matter in source water and distribution systems to form potential DBPs. Many of these DBPs have been shown to cause cancer and reproductive and developmental effects in laboratory animals. More than 200 million people consume water that has been disinfected. Because of the large population exposed, health risks associated with DBPs, even if small, need to be taken seriously.

9.4.1.3 Existing Regulations

- *Microbial contaminants:* the Surface Water Treatment Rule, promulgated in 1989, applies to all public water systems using surface water sources or groundwater sources under the direct influence of surface water. It establishes maximum contaminant level goals (MCLGs) for viruses, bacteria, and *Giardia lamblia.* It also includes treatment technique requirements for filtered and unfiltered systems specifically designed to protect against the adverse health effects of exposure to these microbial pathogens. The Total Coliform Rule, revised in 1989, applies to all PWSs and establishes a maximum contaminant level (MCL) for total coliforms.
- *Disinfection by-products:* in 1979, the USEPA set an interim MCL for total trihalomethanes of 0.10 mg/L as an annual average. This applies to any community water system serving at least 10,000 people that adds a disinfectant to the drinking water during any part of the treatment process.

9.4.1.4 Information Collection Rule

To support the M-DBP rulemaking process, the Information Collection Rule establishes monitoring and data reporting requirements for large public water systems (PWSs) serving at least 100,000 people. This rule is intended to provide the USEPA with information on the occurrence in drinking water of microbial pathogens and DBPs. The USEPA is collecting engineering data on how PWSs currently control such contaminants as part of the Information Collection Rule.

9.4.1.5 Interim Enhanced Surface Water Treatment Rule

The USEPA finalized the Interim Enhanced Surface Water Treatment Rule and Stage 1 Disinfectants and Disinfection By-products Rule in November 1998, as required by the 1996 Amendments to the Safe Drinking Water Act, Section 1412(b)(2)(C). The final rules resulted from formal regulatory negotiations with a wide range of stakeholders that took place in 1992/1993 and 1997.

The Interim Enhanced Surface Water Treatment Rule applies to systems using surface water or groundwater under the direct influence of surface water that serves 10,000 or more persons. The rule also includes provisions for states to conduct sanitary surveys for surface water systems regardless of system size. The rule builds upon the treatment technique requirements of the Surface Water Treatment Rule with the following key additions and modifications:

- maximum contaminant level goal (MCLG) of zero for *Cryptosporidium*
- 2-log *Cryptosporidium* removal requirements for systems that filter
- strengthened combined filter effluent turbidity performance standards
- individual filter turbidity monitoring provisions
- disinfection profiling and benchmarking provisions
- systems using groundwater under the direct influence of surface water now subject to the new rules dealing with *Cryptosporidium*
- inclusion of *Cryptosporidium* in the watershed control requirements for unfiltered public water systems
- requirements for covers on new finished water reservoirs
- sanitary surveys, conducted by states, for all surface water systems regardless of size

The Interim Enhanced Surface Water Treatment Rule, with tightened turbidity performance criteria and required individual filter monitoring, is designed to optimize treatment reliability and to enhance physical removal efficiencies to minimize the *Cryptosporidium* levels in finished water. The rule also includes disinfection benchmark provisions to assure continued levels of microbial protection while facilities take the necessary steps to comply with new DBP standards.

9.4.1.6 Stage 1 Disinfectants and Disinfection By-products Rule

The final Stage 1 Disinfectants and Disinfection By-products Rule applies to community water systems and nontransient noncommunity systems (including those serving fewer than 10,000 people) that add a disinfectant to the drinking water during any part of the treatment process.

The final Stage 1 Disinfectants and Disinfection By-products Rule includes the following key provisions:

- maximum residual disinfectant level goals (MRDLGs) for chlorine (4 mg/L), chloramines (4 mg/L), and chlorine dioxide (0.8 mg/L)
- maximum contaminant level goals (MCLGs) for four trihalomethanes [chloroform (zero), bromodichloromethane (zero), dibromochloromethane (0.06 mg/L), and bromoform (zero)], trichloroacetic acid (0.3 mg/L), bromate (zero), and chlorite (0.8 mg/L)

- MRDLs for three disinfectants: chlorine (4.0 mg/L), chloramines (4.0 mg/L), and chlorine dioxide (0.8 mg/L)
- MCLs for: total trihalomethanes—a sum of the four listed above (0.080 mg/L); haloacetic acids (HAAS) (0.060 mg/L)—a sum of the two listed above plus monochloroacetic acid and mono- and dibromoacetic acids); and two inorganic disinfection by-products [chlorite (1.0 mg/L) and bromate (0.010 mg/L)]
- a treatment technique for removal of DBP precursor material

The terms MRDLG and MRDL (not included in the SDWA) were created during the negotiations to distinguish disinfectants (because of their beneficial use) from contaminants. The final rule includes monitoring, reporting, and public notification requirements for these compounds. This final rule also describes the best available technology (BAT) upon which the MRDLs and MCLs are based.

Future M-DBP Rules include: Long Term 1 and 2 Enhanced Surface Water Treatment Rules, Stage 2 Disinfection By-products Rule, Groundwater Rule, and Filter Backwash Recycling.

9.4.1.6.1 Long Term 1 Enhanced Surface Water Treatment Rule

While the Stage 1 Disinfectants and Disinfection By-products Rule will apply to systems of all sizes, the Interim Enhanced Surface Water Treatment Rule only applies to systems serving 10,000 or more people. A Long Term 1 Enhanced Surface Water Treatment Rule (due in the fall of 2000) will strengthen microbial controls for small systems (i.e., those systems serving fewer than 10,000 people). The rule will also prevent significant increase in microbial risk where small systems take steps to implement the Stage 1 Disinfectants and Disinfection By-products Rule.

The USEPA believes that the rule will generally track the approaches in the Interim Enhanced Surface Treatment Rule for improved turbidity control, including individual filter monitoring and reporting. The rule will also address disinfection profiling and benchmarking. The agency is considering what modifications of some large system requirements may be appropriate for small systems.

9.4.1.6.2 Long Term 2 Enhanced Surface Water Treatment Rule

The SWDA (as amended in 1996) requires the USEPA to finalize a Stage 2 Disinfectants and Disinfection By-products Rule by May 2002. Although the 1996 Amendments do not require the USEPA to finalize a Long Term 2 Enhanced Surface Water Treatment Rule along with the Stage 2 Disinfectants and Disinfection By-products Rule, the USEPA believes that finalizing these rules together to ensure a proper balance between microbial and DBP risks is important.

The USEPA began discussions with stakeholders in December 1998 on the direction for these rules, and anticipates putting the proposed rules into place in early 2001. The intent of the rules is to provide additional public health protection (if needed) from DBPs and microbial pathogens.

9.4.1.6.3 Groundwater Rule

The USEPA is developing a groundwater rule that specifies the appropriate use of disinfection and, equally importantly, addresses other components of groundwater systems to ensure public health protection. More than 158,000 public or community systems serve almost 89 million people through groundwater systems. Ninety-nine percent (157,000) of groundwater systems serve fewer than 10,000 people. However, systems serving more than 10,000 people serve 55%—more than 60 million—of all people who get their drinking water from public groundwater systems. The Groundwater Rule will be promulgated November 2000.

9.4.1.6.4 Filter Backwash Recycling

The 1996 SDWA Amendments require that the USEPA sets a standard on recycling filter backwash within the treatment process of public water systems by August 2000. The regulation will apply to all public water systems, regardless of size. The USEPA is currently gathering data, reviewing literature, and consulting with industry representatives, members of the environmental community, and consulting engineers to identify engineering and cost issues that are salient to regulatory development.

9.4.2 OPPORTUNITIES FOR PUBLIC INVOLVEMENT

The USEPA encourages public input into regulation development. Public meetings and opportunities for public comment on M-DBP rules are announced in the *Federal Register.* EPA's Office of Groundwater and Drinking Water also provides this information for the M-DBP rule and other programs in its online Calendar of Events (www.epa.gov).

9.4.3 FLOCCULANTS

In addition to chlorine and sometimes fluoride, water treatment plants often add several other chemicals, including flocculants, to improve the efficiency of the treatment process; and they all add to the cocktail mix. *Flocculants* are chemical substances added to water to make particles clump together, which improves the effectiveness of filtration. Some of the most common flocculants are polyelectrolytes (polymers)—chemicals with constituents that cause cancer and birth defects and are banned for use by several countries. Although the USEPA classifies them as "probable human carcinogens," it still allows their continued use. Acrylamide and epichlorohydrin are two flocculants used in the United States that are known to be associated with probable cancer risk (Lewis, 1996).

9.5 GROUNDWATER CONTAMINATION

Note that groundwater under the direct influence of surface water comes under the same monitoring regulations as does surface water (i.e., all water open to the atmosphere and subject to surface runoff). The legal definition of *groundwater under the direct influence of surface water* is any water beneath the surface of the ground with (1) significant occurrence of insects or microorganisms, algae, or large diameter pathogens such as *Giardia lamblia*, or (2) significant and relatively rapid shifts in water characteristics such as turbidity, temperature, conductivity, or pH, which closely correlate to climatological or surface water conditions. Direct influence must be determined for individual sources in accordance with criteria established by the state, and determination may be based on site-specific measurements of water quality and/or documentation of well construction characteristics and geology with field evaluation.

Generally, most groundwater supplies in the U.S. are of good quality and produce essential quantities. The full magnitude of groundwater contamination in the U.S. is, however, not fully documented, and federal, state, and local efforts continue to assess and address the problems (Rail, 1985).

Groundwater supplies about 25% of the freshwater used for all purposes in the United States, including irrigation, industrial uses, and drinking water (about 50% of the U.S. population relies on groundwater for drinking water). John Chilton (1998) points out that the groundwater aquifers beneath or close to Mexico City provide the area with more than 3.2 billion liters per day. But, as groundwater pumping increases to meet water demand, it can exceed the aquifers' rates of replenishment, and in many urban aquifers, water levels show long-term decline. With excessive extraction comes a variety of other undesirable effects, including:

- increased pumping costs
- changes in hydraulic pressure and underground flow directions (in coastal areas, this results in seawater intrusion)
- saline water drawn up from deeper geological formations
- poor-quality water from polluted shallow aquifers leaking downwards

Severe depletion of groundwater resources is often compounded by a serious deterioration in its quality. Thus, without a doubt, the contamination of a groundwater supply should be a concern of those drinking water practitioners responsible for supplying a community with potable water provided by groundwater.

Despite our strong reliance on groundwater, groundwater has for many years been one of the most neglected natural resources. Groundwater has been ignored because it is less visible than other environmental resources—rivers or lakes, for example. What the public cannot see or observe, the public doesn't worry about, or even think about. However, recent publicity about events concerning groundwater contamination is making the public more aware of the problem, and the regulators have also taken notice.

Are natural contaminants a threat to human health, harbingers of serious groundwater pollution events? No, not really. The main problem with respect to serious groundwater pollution has been human activities. When we (all of us) improperly dispose of wastes, or spill hazardous substances onto/into the ground, we threaten groundwater, and in turn, the threat passes on to public health.

Let's take a closer look at a few sources of groundwater contamination.

9.5.1 SOURCES OF GROUNDWATER CONTAMINATION

Several sources of groundwater contamination present cause for concern for the drinking water practitioner. Why? Consider the importance of groundwater. People depend upon groundwater in every state, and its usage accounts for approximately one-fourth of all water used. This consumption includes about 35% of water withdrawn for municipal water supplies.

9.5.1.1 Underground Storage Tanks (USTs)

If we looked at a map of the United States marked with the exact location of every underground storage tank (UST) (we wish such a map existed!), with the exception of isolated areas, most of us would be surprised at the large number of tanks buried underground. With so many buried tanks, it should come as no surprise that structural failures arising from a wide variety of causes have occurred over the years. Subsequent leaking has become a huge source of contamination that affects the quality of local groundwaters.

Note: A UST is any tank, including any underground piping connected to the tank, that has at least 10% of its volume below ground (USEPA, 1987).

The fact is, leakage of petroleum and its products from USTs occurs more often than we generally realize. This widespread problem has been a major concern and priority in the U.S. for well over a decade. In 1987, the USEPA promulgated regulations for many of the nation's USTs, and much progress has been made in mitigating this problem to date.

When a UST leak or past leak is discovered, the contaminants released to the soil and thus to groundwater might seem, for the average person, straightforward to identify: fuel oil, diesel, and gasoline. However, even though true that these are the most common contaminants released from leaking USTs, others also present problems. For example, in the following section, we discuss one such contaminant, a by-product of gasoline that is not well known, to help illustrate the magnitude of leaking USTs.

9.5.1.1.1 MtBE

In December 1997, the USEPA issued a drinking water advisory titled: Consumer Acceptability Advice and Health Effects Analysis on Methyl Tertiary-Butyl Ether (MtBE). The purpose of the advisory was to provide guidance to communities exposed to drinking water contaminated with MtBE.

Note: A USEPA advisory is usually initiated to provide information and guidance to individuals or agencies concerned with potential risk from drinking water contaminants for which no national regulations currently exist. Advisories are not mandatory standards for action, and are used only for guidance. They are not legally enforceable, and are subject to revision as new information becomes available. The USEPA's health advisory program is recognized in the Safe Drinking Water Act Amendments of 1996, which state in section 102(b)(1)(F): "The Administrator may publish health advisories (which are not regulations) or take other appropriate actions for contaminants not subject to any national primary drinking water regulation."

As its title indicates, this advisory for MtBE includes consumer acceptability advice as "appropriate" under this statutory provision, as well as a health effects analysis.

9.5.1.1.1.1 WHAT IS MtBE?

MtBE is a volatile, organic chemical. Since the late 1970s, MtBE has been used as an octane enhancer in gasoline. Because it promotes more complete burning of gasoline (thereby reducing carbon monoxide and ozone levels) it is commonly used as a gasoline additive in localities that do not meet the National Ambient Air Quality Standards.

In the Clean Air Act of 1990, Congress mandated the use of reformulated gasoline (RFG) in areas of the country with the worst ozone or smog problems. RFG must meet certain technical specifications set forth in the Act, including a specific oxygen content. Ethanol and MtBE are the primary oxygenates used to meet the oxygen content requirement. MtBE is used in about 84% of RFG supplies. Currently, 32 areas in a total of 18 states are participating in the RFG program, and RFG accounts for about 30% of gasoline nationwide.

Studies identify significant air quality and public health benefits that directly result from the use of fuels oxygenated with MtBE, ethanol, or other chemicals. The refiners' 1995/1996 fuel data submitted to the USEPA indicates that the national emissions benefits exceeded those required. The 1996 Air Quality Trends Report shows that toxic air pollutants declined significantly between 1994 and 1995. Early analysis indicates this progress may be attributable to the use of RFG. Starting in the year 2000, required emission reductions are substantially greater, at about 27% for volatile organic compounds, 22% for toxic air pollutants, and 7% for nitrogen oxides.

Note: When gasoline that has been oxygenated with MtBE comes in contact with water, large amounts of MtBE dissolve. At 25°C, the water solubility of MtBE is about 5000 milligrams per liter for a gasoline that is 10% MtBE by weight. In contrast, for a nonoxygenated gasoline, the total hydrocarbon solubility in water is typically about 120 milligrams per liter. MtBE sorbs only weakly to soil and aquifer material; therefore, sorption will not significantly retard MtBE's transport by groundwater. In addition, the compound generally resists degradation in groundwater (Squillace et al., 1998).

9.5.1.1.1.2 WHY IS MtBE A DRINKING WATER CONCERN?

A limited number of instances of significant contamination of drinking water with MtBE have occurred because of leaks from underground and above ground petroleum storage tank systems and pipelines. Due to its small molecular size and solubility in water, MtBE moves rapidly into groundwater, faster than do other constituents of gasoline. Public and private wells have been contaminated in this manner. Nonpoint sources (such as recreational watercraft) are most likely to be the cause of small amounts of contamination in a large number of shallow aquifers and surface waters. Air deposition through precipitation of industrial or vehicular emissions may also contribute to surface water contamination. The extent of any potential for buildup in the environment from such deposition is uncertain.

9.5.1.1.1.3 IS MtBE IN DRINKING WATER HARMFUL?

Based on the limited sampling data currently available, most concentrations at which MtBE has been found in drinking water sources are unlikely to cause adverse health effects. However, the USEPA is continuing to evaluate the available information and is doing additional research to seek more definitive estimates of potential risks to humans from drinking water.

There are no data on the effects on humans of drinking MtBE-contaminated water. In laboratory tests on animals, cancer and noncancer effects occur at high levels of exposure. These tests were conducted by inhalation exposure or by introducing the chemical in oil directly to the stomach. The tests support a concern for potential human hazard. Because the animals were not exposed through drinking water, significant uncertainties exist concerning the degree of risk associated with human exposure to low concentrations typically found in drinking water.

9.5.1.1.1.4 HOW CAN PEOPLE BE PROTECTED?

MtBE has a very unpleasant taste and odor, and these properties make contaminated drinking water unacceptable to the public. The advisory recommends control levels for taste and odor acceptability that will also protect against potential health effects.

Studies conducted on the concentrations of MtBE in drinking water determined the level at which individuals can detect the odor or taste of the chemical. Humans vary widely in the concentrations they are able to detect. Some who are sensitive can detect very low concentrations. Others do not taste or smell the chemical, even at much higher concentrations. The presence or absence of other natural or water treatment chemicals sometimes masks or reveals the taste or odor effects.

Studies to date have not been extensive enough to completely describe the extent of this variability, or to establish a population response threshold. Nevertheless, we conclude from the available studies that keeping concentrations in the range of 20 to 40 micrograms per liter (μg/L) of water or below will likely avert unpleasant taste and odor effects, recognizing that some people may detect the chemical below this.

Concentrations in the range of 20 to 40 μg/L are about 20,000 to 100,000 (or more) times lower than the range of exposure levels in which cancer or noncancer effects were observed in rodent tests. This margin of exposure lies within the range of margins of exposure typically provided to protect against cancer effects by the National Primary Drinking Water Standards under the Federal Safe Drinking Water Act—a margin greater than such standards typically provided to protect against noncancer effects. Protection of the water source from unpleasant taste and odor as recommended also protects consumers from potential health effects.

The USEPA also notes that occurrences of groundwater contamination observed at or above this 20–40 μg/L taste and odor threshold—that is, contamination at levels that may create consumer acceptability problems for water supplies—have, to date, resulted from leaks in petroleum storage tanks or pipelines, not from other sources.

9.5.1.1.1.5 RECOMMENDATIONS FOR STATE OR PUBLIC WATER SUPPLIERS

Public water systems that conduct routine monitoring for volatile organic chemicals can test for MtBE at little additional cost, and some states are already moving in this direction.

Public water systems detecting MtBE in their source water at problematic concentrations can remove MtBE from water using the same conventional treatment techniques that are used to clean up other contaminants originating from gasoline releases—air stripping and granular activated carbon (GAC), for example. However, because MtBE is more soluble in water and more resistant to biodegradation than other chemical constituents in gasoline, air stripping and GAC treatment require additional optimization, and must often be used together to effectively remove MtBE from water. The costs of removing MtBE are higher than when treating for gasoline releases that do not contain MtBE. Oxidation of MtBE using UV/peroxide/ozone treatment may also be feasible, but typically has higher capital and operating costs than air stripping and GAC.

Note: Of the 60 volatile organic compounds (VOCs) analyzed in samples of shallow ambient groundwater collected from eight urban areas during 1993–1994 as part of the U.S. Geological Survey's National Water Quality Assessment program, MtBe was the second most frequently detected compound [after trichloromethane (chloroform)] (Squillace et al., 1998).

9.5.2 INDUSTRIAL WASTES

Since industrial waste represents a significant source of groundwater contamination, drinking water practitioners and others expend an increasing amount of time in abating or mitigating pollution events that damage groundwater supplies.

Groundwater contamination from industrial wastes usually begins with the practice of disposing of industrial chemical wastes in surface impoundments—unlined landfills or lagoons, for example. Fortunately, these practices, for the most part, are part of our past. Today, we know better. We now know that what is most expedient or least expensive does not work for industrial waste disposal practices. We have found through actual experience that in the long run, just the opposite is true. With respect to health hazards and the costs of cleanup activities, ensuring clean or unpolluted groundwater supplies is very expensive, and utterly necessary.

9.5.3 SEPTIC TANKS

Septage from septic tanks is a biodegradable waste capable of affecting

the environment through water and air pollution. The potential environmental problems associated with use of septic tanks are magnified when you
consider that subsurface sewage disposal systems (septic tanks) are used by
almost one-third of the United States population.

Briefly, a septic tank and leaching field system traps and stores solids
while the liquid effluent from the tank flows into a leaching or absorption
field, where it slowly seeps into the soil and degrades naturally.

The problem with subsurface sewage disposal systems such as septic tanks
is that most of the billions of gallons of sewage that enter the ground each
year are not properly treated. Because of faulty construction or lack of
maintenance, not all of these systems work properly.

Experience has shown that septic disposal systems are frequently sources
of fecal bacteria and virus contamination of water supplies taken from
private wells. Many septic tank owners dispose of detergents, nitrates,
chlorides, and solvents in their septic systems, or use solvents to treat their
sewage waste. A septic tank cleaning fluid that is commonly used contains
organic solvents (trichlorethylene or TCE), potential human carcinogens
that in turn pollute the groundwater in areas served by septic systems.

9.5.4 LANDFILLS

Humans have been disposing of waste by burying it in the ground since
time immemorial. In the past, this practice was largely uncontrolled, and the
disposal sites (i.e., garbage dumps) were places where municipal solid
wastes were simply dumped on and into the ground without much thought
or concern. Even in this modern age, landfills have been used to dispose of
trash and waste products at controlled locations that are then sealed and
buried under earth. Now such practices are increasingly seen as a less than
satisfactory disposal method, because of the long-term environmental impact of waste materials in the ground and groundwater.

Unfortunately, many of the older (and even some of the newer) sites were
located in low-lying areas with high groundwater tables. *Leachate* (seepage
of liquid through the waste), high in BOD, chloride, organics, heavy metals,
nitrate, and other contaminants, has little difficulty reaching the groundwater
in such disposal sites. In the U.S., literally thousands of inactive or abandoned dumps like this exist.

9.5.5 AGRICULTURE

Fertilizers and pesticides are the two most significant groundwater contaminants that result from agricultural activities. The impact of agricultural
practices wherein fertilizers and pesticides are normally used is dependent
upon local soil conditions. If, for example, the soil is sandy, nitrates from

fertilizers are easily carried through the porous soil into the groundwater, contaminating private wells.

Pesticide contamination of groundwater is a subject of national importance because groundwater is used for drinking water by about 50% of the nation's population. This especially concerns people living in the agricultural areas where pesticides are most often used, as about 95% of that population relies upon groundwater for drinking water. Before the mid-1970s, the common thought was that soil acted as a protective filter, one that stopped pesticides from reaching groundwater. Studies have now shown that this is not the case. Pesticides can reach water-bearing aquifers below ground from applications onto crop fields, seepage of contaminated surface water, accidental spills and leaks, improper disposal, and even through injection of waste material into wells.

Pesticides are mostly modern chemicals. Many hundreds of these compounds are used, and extensive tests and studies of their effect on humans have not been completed. That leads us to wonder just how concerned we should be about their presence in our drinking water. Certainly, treating pesticides as potentially dangerous and thus handling them with care would be wise. We can say they pose a potential danger if they are consumed in large quantities, but as any experienced scientist knows, you cannot draw factual conclusions unless scientific tests have been done. Some pesticides have had a designated Maximum Contaminant Limit (MCL) in drinking water set by the USEPA, but many have not. Another serious point to consider is the potential effect of combining more than one pesticide in drinking water, which might be different than the effects of each individual pesticide alone. This is another situation where we don't have sufficient scientific data to draw reliable conclusions.

9.5.6 SALT WATER INTRUSION

In many coastal cities and towns as well as in island locations, the intrusion of salty seawater presents a serious water-quality problem. Because freshwater is lighter than salt water (the specific gravity of seawater is about 1.025), it will usually float above a layer of salt water. When an aquifer in a coastal area is pumped, the original equilibrium is disturbed and salt water replaces the freshwater (Viessman and Hammer, 1998). The problem is compounded by increasing population, urbanization, and industrialization, which increase use of groundwater supplies. In such areas, while groundwater is heavily drawn upon, the quantity of natural groundwater recharge is decreased because of the construction of roads, tarmac, and parking lots, which prevent rainwater from infiltrating, decreasing the groundwater table elevation.

In coastal areas, the natural interface between the fresh groundwater flowing from upland areas and the saline water from the sea is constantly under attack by human activities. Since seawater is approximately 2.5 times more dense than freshwater, a high pressure head of seawater occurs (in relationship to freshwater), which results in a significant rise in the seawater boundary. Potable water wells close to this rise in sea level may have to be abandoned because of salt water intrusion.

9.5.7 OTHER SOURCES OF GROUNDWATER CONTAMINATION

To this point, we have discussed only a few of the many sources of groundwater contamination. For example, we have not discussed mining and petroleum activities that lead to contamination of groundwater, or contamination caused by activities in urban areas. Both of these are important sources. Urban activities (including spreading salt on roads to keep them ice-free during winter) eventually contribute to contamination of groundwater supplies. Underground injection wells used to dispose of hazardous materials can lead to groundwater contamination. As we've discussed, underground storage tanks (USTs) are also significant contributors to groundwater pollution.

Other sources of groundwater contamination include these items on our short list:

- waste tailings
- residential disposal
- urban runoff
- hog wastes
- biosolids
- land applied wastewater
- graveyards
- de-icing salts
- surface impoundments
- waste piles
- animal feeding operations
- natural leaching
- animal burial
- mine drainage
- pipelines
- open dumps
- open burning
- atmospheric pollutants

We did not list raw sewage, because, for the most part, raw sewage is no longer routinely dumped into our nation's wells or into our soil. Sewage treatment plants effectively treat wastewater so that it can be safely discharged to local waterbodies. In fact, the amount of pollution being discharged from these plants has been cut by over one-third during the past 20 years, even as the number of people served has doubled.

Yet in some areas, raw sewage spills still occur, sometimes because an underground sewer line is blocked, broken, or too small, or because periods of heavy rainfall overload the capacity of the sewer line or sewage treatment plant, so that overflows into city streets or streams occur. Some of this sewage finds its way to groundwater supplies.

9.6 SUMMARY

The best way to prevent groundwater pollution is to stop it from occurring in the first place. Unfortunately, a perception held by many is that natural purification of chemically contaminated ground can take place on its own, without the aid of human intervention. To a degree this is true; however, natural purification functions on its own time, not on human time. Natural purification could take decades and perhaps centuries. The alternative? Remediation. But remediation and mitigation don't come cheap. When groundwater is contaminated, the cleanup efforts are sometimes much too expensive to be practical.

The USEPA has established the Groundwater Guardian program. The Program is a voluntary way to improve drinking water safety. Established and managed by a nonprofit organization in the Midwest and strongly promoted by the USEPA, this program focuses on communities that rely on groundwater for their drinking water. It provides special recognition and technical assistance to help communities protect their groundwater from contamination. Since beginning in 1994, Groundwater Guardian programs have been established in nearly 100 communities in 31 states (USEPA, 1996).

9.7 REFERENCES

Chilton, J., *Dry or Drowning.* Chicago: Beson, 1998.

Harr, J., *A Civil Action.* New York: Vintage Books, 1995.

Hill, M. K., *Understanding Environmental Pollution.* Cambridge, UK: Cambridge University Press, 1997.

Kay, J., "Chemicals Used to Cleanse Water Can Also Cause Problems." *San Francisco Examiner*, October 3, 1996.

Lewis, S. A., *The Sierra Club Guide to Safe Drinking Water.* San Francisco: Sierra Club Books, 1996.

Outwater, A., *Water: A Natural History.* New York: John Wiley and Sons, pp. 175–176, 1996.

Rail, C. D., "Groundwater Monitoring within an Aquifer—A Protocol." *Journal of Environmental Health.* 48(3):128–132, 1985.

Squillace, P. J., Pankow, J. F., Korte, N. E., and Zogorski, J. S., "Environmental Behavior and Fate of Methyl Tertiary-Butyl Ether." In *Water Online,* @www.wateronline.com, November 4, 1998.

USA Today, (Schlichtmann, Jan) Feb. 4, 1999.

USEPA. *Proposed Regulations for Underground Storage Tanks: What's in the Pipeline?* Washington, D.C.: Office of Underground Storage Tanks, 1987.

USEPA. *Targeting High Priority Problems,* @epamail.epa.gov, 1996.

USEPA. *Drinking Water Priority Rulemaking: Microbial and Disinfection By-products Rules.* Washington, D.C.: United States Environmental Protection Agency, EPA 815-F-95-0014, 1998.

Viessman, W., Jr. and Hammer, M. J., *Water Supply and Pollution Control*, 6th Edition. Menlo Park, CA: Addison-Wesley, 1998.

CHAPTER 10

Drinking Water Monitoring

It is not enough that you should understand about applied science in order that your work may increase man's blessings. Concern for the man himself and his fate must always form the chief interest of all technical endeavors; concern for the great unsolved problems of the organization of labor and the distribution of goods in order that the creations of our mind shall be a blessing and not a curse to mankind.

Never forget this in the midst of your diagrams and equations. (Albert Einstein, 1931)

10.1 INTRODUCTION

WHEN we speak of drinking water monitoring, we refer to water quality monitoring based on three criteria: (1) to ensure to the extent possible that the water is not a danger to public health; (2) to ensure that the water provided at the tap is as aesthetically pleasing as possible; and (3) to ensure compliance with applicable regulations. To meet these goals, all public water systems must monitor water quality to some extent. The degree of monitoring employed is dependent on local needs and requirements, and on the type of water system; small water systems using good-quality water from deep wells may only need to provide occasional monitoring, but systems using surface water sources must test water quality frequently (AWWA, 1995).

Drinking water must be monitored to provide adequate control of the entire water drawing/treatment/conveyance system. *Adequate control* is defined as monitoring employed to assess the present level of water quality, so action can be taken to maintain the required level (whatever that might be).

We define *water quality monitoring* as the sampling and analysis of water constituents and conditions. When we monitor, we collect data. As a monitoring program is developed, deciding the reasons for collecting the information is important. The reasons are defined by establishing a set of objectives, which includes a description of who will collect the information.

It may surprise you to know that today the majority of people collecting data are volunteers, not necessarily professional drinking water practitioners. These volunteers have a vested interest in their local stream, lake, or other body of water, and in many cases are proving they can successfully carry out a water quality monitoring program.

10.2 IS THE WATER GOOD OR BAD?[13]

To answer the question "Is the water good or bad?" we must consider two factors. First, we return to the basic principles of water quality monitoring—sampling and analyzing water constituents and conditions. These constituents may include:

- introduced pollutants, such as pesticides, metals, and oil
- constituents found naturally in water that can nevertheless be affected by human sources, such as dissolved oxygen, bacteria, and nutrients

The magnitude of their effects is influenced by properties such as pH and temperature. Temperature influences the quantity of dissolved oxygen that water is able to contain, and pH affects the toxicity of ammonia, for example.

Let's get back to whether water is "good" or "bad." The second factor to be considered is that the only valid way to answer this question is to conduct tests that must be compared to some form of water quality standards. If simply assigning a "good" and "bad" value to each test factor were possible, the meters and measuring devices in water quality test kits would be much easier to make. Instead of fine graduations, they could simply have a "good" and a "bad" zone.

Water quality—the difference between "good" and "bad" water—must be interpreted according to the intended use of the water. For example, the perfect balance of water chemistry that assures a sparkling clear, sanitary swimming pool would not be acceptable as drinking water and would be a deadly environment for many biota. Consider Table 10.1.

In another example, widely different levels of fecal coliform bacteria are considered acceptable, depending on the intended use of the water.

TABLE 10.1. Total Residual Chlorine (TRC) mg/L.

0.06	Toxic to striped bass larvae
0.31	Toxic to white perch larvae
0.5 to 1.0	Typical drinking water residual
1.0 to 3.0	Recommended for swimming pools

[13]Much of the information presented in the following sections is from USEPA's 2.841B97003 *Volunteer Stream Monitoring: A Methods Manual*, 1997.

TABLE 10.2. Fecal Coliform Bacteria per 100 mL of Water.

Desirable	Permissible	Type of Water Use
0	0	Potable and well water (for drinking)
<200	<1000	Primary contact water (for swimming)
<1000	<5000	Secondary contact water (for boating and fishing)

State and local water quality practitioners as well as volunteers have been monitoring water quality conditions for many years. In fact, until the past decade or so (until biological monitoring protocols were developed and began to take hold), water quality monitoring was generally considered the primary way of identifying water pollution problems. Today, professional water quality practitioners and volunteer program coordinators alike are moving toward approaches that combine chemical, physical, and biological monitoring methods to achieve the best picture of water quality conditions.

Water quality monitoring can be used for many purposes:

(1) *To identify whether waters are meeting designated uses.* All states have established specific criteria (limits on pollutants) identifying what concentrations of chemical pollutants are allowable in their waters. When chemical pollutants exceed maximum or minimum allowable concentrations, waters may no longer be able to support the beneficial uses—such as fishing, swimming, and drinking—for which they have been designated (see Table 10.2). Designated or intended uses and the specific criteria that protect them (along with antidegradation statements that say waters should not be allowed to deteriorate below existing or anticipated uses) together form water quality standards. State water quality professionals assess water quality by comparing the concentrations of chemical pollutants found in streams to the criteria in the state's standards, and so judge whether streams are meeting their designated uses.

Water quality monitoring, however, might be inadequate for determining whether aquatic life needs are being met in a stream. While some constituents (such as dissolved oxygen and temperature) are important to maintaining healthy fish and aquatic insect populations, other factors (such as the physical structure of the stream and the condition of the habitat) play an equal or greater role. Biological monitoring methods are generally better suited to determining whether or not aquatic life is supported.

(2) *To identify specific pollutants and sources of pollution.* Water quality monitoring helps link sources of pollution to stream quality problems because it identifies specific problem pollutants. Since certain activities

tend to generate certain pollutants (bacteria and nutrients are more likely to come from an animal feedlot than an automotive repair shop), a tentative link to what would warrant further investigation or monitoring can be formed.

(3) *To determine trends.* Chemical constituents that are properly monitored (i.e., using consistent time of day and on a regular basis using consistent methods) can be analyzed for trends over time.

(4) *To screen for impairment.* Finding excessive levels of one or more chemical constituents can serve as an early warning "screen" for potential pollution problems.

10.3 STATE WATER QUALITY STANDARDS PROGRAMS

Each state has a program to set standards for the protection of each body of water within its boundaries. Standards for each body of water are developed that:

- depend on the water's designated use (Table 10.1)
- are based on USEPA national water quality criteria and other scientific research into the effects of specific pollutants on different types of aquatic life and on human health
- may include limits based on the biological diversity of the body of water (the presence of food and prey species)

State water quality standards set limits on pollutants and establish water quality levels that must be maintained for each type of water body, based on its designated use.

Resources for this type of information include:

- USEPA Water Quality Criteria Program
- U.S. Fish and Wildlife Service Habitat Suitability Index Models (for specific species of local interest)

Your own monitoring test results can be plotted against these standards to provide a focused, relevant, required assessment of water quality.

10.4 DESIGNING A WATER QUALITY MONITORING PROGRAM

The first step in designing a water quality monitoring program is to determine the purpose for the monitoring. This aids in selection of parameters to monitor. This decision should be based on factors, including:

- types of water quality problems and pollution sources that will likely be encountered (see Table 10.3)

TABLE 10.3. Water Quality Problems and Pollution Sources.

Source	Common Associated Chemical Pollutants
Cropland	Turbidity, phosphorus, nitrates, temp., total solids
Forestry harvest	Turbidity, temp., total solids
Grazing land	Fecal bacteria, turbidity, phosphorus
Industrial discharge	Temp., conductivity, total solids, toxics, pH
Mining	pH, alkalinity, total dissolved solids
Septic systems	Fecal bacteria, (i.e., *Escherichia coli, enterococcus*), nitrates, phosphorus, dissolved oxygen/biochemical oxygen demand, conductivity, temp.
Sewage treatment plants	Dissolved oxygen and BOD, turbidity, conductivity, phosphorus, nitrates, fecal bacteria, temp., total solids, pH
Construction	Turbidity, temp., dissolved oxygen and BOD, total solids, toxics
Urban runoff	Turbidity, phosphorus, nitrates, temp., conductivity, dissolved oxygen, BOD

- cost of available monitoring equipment
- precision and accuracy of available monitoring equipment
- capabilities of the monitors

We discuss the parameters most commonly monitored by drinking water practitioners in streams (i.e., we assume, for illustration and discussion purposes, that our water source is a surface water stream) in detail in this chapter. They include dissolved oxygen, biochemical oxygen demand, temperature, pH, turbidity, total orthophosphate, nitrates, total solids, conductivity, total alkalinity, fecal bacteria, apparent color, odor, and hardness (see Figure 10.1).

Note: When monitoring water supplies under the Safe Drinking Water Act (SDWA) or the National Pollutant Discharge Elimination System (NPDES), utilities must follow test procedures approved by the USEPA for these purposes. Additional testing requirements under these and other federal programs are published as amendments in the Federal Register.

Except when monitoring discharges for specific compliance purposes, a large number of approximate measurements can provide more useful information than one or two accurate analyses. Since water quality and chemistry continually change, making periodic, repetitive measurements and observations that indicate the range of water quality is necessary, rather than testing the quality at any single moment. The more complex a water system, the more time is required to observe, understand, and draw conclusions regarding the cause and effect of changes in the particular system.

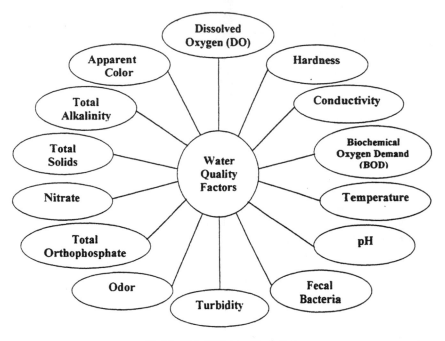

Figure 10.1 Water quality factors.

10.5 GENERAL PREPARATION AND SAMPLING CONSIDERATIONS

The sections that follow detail specific equipment considerations and analytical procedures for each of the most common water quality parameters. However, two general tasks should be accomplished any time water samples are taken. We discuss them below.

10.5.1 PREPARATION OF SAMPLING CONTAINERS

Sampling devices should be corrosion resistant, easily cleaned, and capable of collecting desired samples safely and in accordance with test requirements. Whenever possible, assign a sampling device to each sampling point. Sampling equipment must be cleaned on a regular schedule to avoid contamination.

Note: Some tests require special equipment to ensure the sample is representative. Dissolved oxygen and fecal bacteria sampling require special equipment and/or procedures to prevent collection of nonrepresentative samples.

Reused sample containers and glassware must be cleaned and rinsed before the first sampling run and after each run by following Method A or Method B described below. The most suitable method depends on the parameter being measured.

10.5.1.1 Method A: General Preparation of Sampling Containers

Use the following method when preparing all sample containers and glassware for monitoring conductivity, total solids, turbidity, pH, and total alkalinity. Wearing latex gloves:

(1) Wash each sample bottle or piece of glassware with a brush and phosphate-free detergent.
(2) Rinse three times with cold tap water.
(3) Rinse three times with distilled or deionized water.

10.5.1.2 Method B: Acid Wash Procedures

Use this method when preparing all sample containers and glassware for monitoring nitrates and phosphorus. Wearing latex gloves:

(1) Wash each sample bottle or piece of glassware with a brush and phosphate-free detergent.
(2) Rinse three times with cold tap water.
(3) Rinse with 10% hydrochloric acid.
(4) Rinse three times with deionized water.

10.5.2 SAMPLE TYPES

Two basic types of samples are commonly used for water quality testing: grab samples and composite samples. The type of sample used depends on the specific test, the reason the sample is being collected, and the applicable regulatory requirements.

Grab samples are collected at one time and one location. They are representative only of the conditions at the time of collection.

Grab samples must be used to determine pH, total residual chlorine (TRC), dissolved oxygen (DO), and fecal coliform concentrations. Grab samples may also be used for any test which does not specifically prohibit their use.

Note: Before collecting samples for any test procedure, it is best to review the sampling requirements of the test.

Composite samples consist of a series of individual grab samples collected over a specified period of time in proportion to flow. The individual grab samples are mixed together in proportion to the flow rate at the time the sample was collected to form the composite sample.

10.5.3 COLLECTING SAMPLES (FROM A STREAM)

In general, sample away from the streambank in the main current. Never sample stagnant water. The outside curve of the stream is often a good place to sample, since the main current tends to hug this bank. In shallow stretches, carefully wade into the center current to collect the sample.

A boat is required for deep sites. Try to maneuver the boat into the center of the main current to collect the water sample.

When collecting a water sample for analysis in the field or at the lab, follow the steps below.

10.5.3.1 Whirl-pak® Bags

(1) Label the bag with the site number, date, and time.

(2) Tear off the top of the bag along the perforation above the wire tab just prior to sampling. Avoid touching the inside of the bag. If you accidentally touch the inside of the bag, use another one.

(3) *Wading.* Try to disturb as little bottom sediment as possible. In any case, be careful not to collect water that contains bottom sediment. Stand facing upstream. Collect the water sample in front of you.

Boat. Carefully reach over the side and collect the water sample on the upstream side of the boat.

(4) Hold the two white pull tabs in each hand and lower the bag into the water on your upstream side with the opening facing upstream. Open the bag midway between the surface and the bottom by pulling the white pull tabs. The bag should begin to fill with water. You may need to "scoop" water into the bag by drawing it through the water upstream and away from you. Fill the bag no more than 3/4 full!

(5) Lift the bag out of the water. Pour out excess water. Pull on the wire tabs to close the bag. Continue holding the wire tabs and flip the bag over at least four to five times quickly to seal the bag. Don't try to squeeze the air out of the top of the bag. Fold the ends of the bag, being careful not to puncture the bag. Twist them together, forming a loop.

(6) Fill in the bag number and/or site number on the appropriate field data sheet. *This is important.* It is the only way the lab specialist will know which bag goes with which site.

(7) If samples are to be analyzed in a lab, place the sample in the cooler with ice or cold packs. Take all samples to the lab.

10.5.3.2 Screw-Cap Bottles

To collect water samples using screw-cap sample bottles, use the following procedures (see Figure 10.2).

1   2

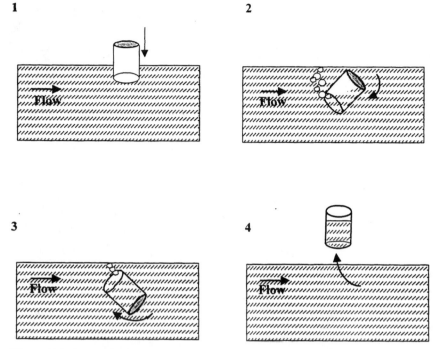

3   4

Figure 10.2 Taking a water sample. Turn the bottle into the current and scoop in an upstream direction.

(1) Label the bottle with the site number, date, and time.

(2) Remove the cap from the bottle just before sampling. Avoid touching the inside of the bottle or the cap. If you accidentally touch the inside of the bottle, use another one.

(3) *Wading.* Try to disturb as little bottom sediment as possible. In any case, be careful not to collect water that has sediment from bottom disturbance. Stand facing upstream. Collect the water sample on your upstream side, in front of you. You may also tape your bottle to an extension pole to sample from deeper water.

Boat. Carefully reach over the side and collect the water sample on the upstream side of the boat.

(4) Hold the bottle near its base and plunge it (opening downward) below the water surface. If you are using an extension pole, remove the cap, turn the bottle upside down, and plunge it into the water, facing upstream. Collect a water sample 8 to 12 inches beneath the surface, or midway between the surface and the bottom if the stream reach is shallow.

(5) Turn your bottle underwater into the current and away from you. In slow-moving stream reaches, push the bottle underneath the surface and away from you in the upstream direction.

(6) Leave a 1-inch air space (except for DO and BOD samples). Do not fill the bottle completely (so that the sample can be shaken just before analysis). Recap the bottle carefully, remembering not to touch the inside.

(7) Fill in the bottle number and/or site number on the appropriate field data sheet. This is important because it tells the lab specialist which bottle goes with which site.

(8) If the samples are to be analyzed in the lab, place them in the cooler for transport to the lab.

10.5.4 SAMPLE PRESERVATION AND STORAGE

Samples can change very rapidly. However, no single preservation method will serve for all samples and constituents. If analysis must be delayed, follow the instructions for sample preservation and storage listed in *Standard Methods,* or those specified by the laboratory that will eventually process the samples (see Table 10.4). In general, handle the sample in a way that prevents changes from biological activity, physical alterations, or chemical reactions. Cool the sample to reduce biological and chemical reactions. Store in darkness to suspend photosynthesis. Fill the sample

TABLE 10.4. Recommended Sample Storage and Preservation Techniques.

Test Factor	Container Type	Preservation	Max. Storage Time Recommended/ Regulatory
Alkalinity	P, G	Refrigerate	24 hr/14 days
BOD	P, G	Refrigerate	6 hr/48 hr
Conductivity	P, G	Refrigerate	28 days/28 days
Hardness	P, G	Lower pH to <2	6 mos/6 mos
Nitrate	P, G	Analyze ASAP	48 hr/48 hr
Nitrite	P, G	Analyze ASAP	none/48 hr
Odor	G	Analyze ASAP	6 hr/NR
Oxygen, dissolved:			
Electrode	G	Immed. anal.	0.5 hr/stat
Winkler	G	"Fix" immed.	8 hr/8 hr
pH	P, G	Immed. anal.	2 hr/stat
Phosphate	G (A)	Filter immed., refrigerate	48 hr/NR
Salinity	G, wax seal	Immed. anal. or use wax seal	6 mos/NR
Temperature	P, G	Immed. anal.	stat/stat
Turbidity	P, G	Analyze same day or store in dark up to 24 hr, refrigerate	24 hr/48 hr

Refrigerate = store at 4°C, in dark; P = plastic (polyethylene or equivalent); G = glass; G (A) = glass rinsed with 1 + 1 HNO_3; Stat = no storage allowed; NR = none recommended.
Source: Adapted from *Standard Methods,* 1989.

container completely to prevent the loss of dissolved gases. Metal cations such as iron and lead and suspended particles may adsorb onto container surfaces during storage.

10.5.5 STANDARDIZATION OF METHODS

References used for sampling and testing must correspond to those listed in the most current Federal Regulation. For the majority of tests, to compare the results of either different water quality monitors or the same monitors over the course of time requires some form of standardization of the methods. The American Public Health Association (APHA) recognized this requirement when in 1899 the Association appointed a committee to draw up standard procedures for the analysis of water. The report (published in 1905) constituted the first edition of what is now known as *Standard Methods for the Examination of Water and Wastewater* or *Standard Methods*. This book is now in its 20th edition and serves as the primary reference for water testing methods, and as the basis for most EPA-approved methods.

Note: Some of the methods used by volunteer monitors are based directly on procedures outlined in the APHA *Standard Methods*. Although many methods used by volunteer monitors are not described in *Standard Methods*, they can be standardized to provide repeatable and comparable data. Instructions, training, and audit procedures all play a role in standardizing the methods used by volunteer monitors.

10.6 TEST METHODS[14]

Descriptions of general methods to help you understand how each works in specific test kits follow. Always use the specific instructions included with the equipment and individual test kits.

Most water analyses are conducted either by titrimetric analyses or colorimetric analyses. Both methods are easy to use and provide accurate results.

10.6.1 TITRIMETRIC

Titrimetric analyses are based on adding a solution of known strength (the titrant) to a specific volume of a treated sample in the presence of an indicator. The indicator produces a color change indicating the reaction is complete. Titrants are generally added by a titrator (microburette) or a precise glass pipette.

[14]Much of the information is adapted from *Standard Methods* and *The Monitor's Handbook*, LaMotte Company, Chestertown, Maryland, 1992.

10.6.2 COLORIMETRIC

Two basic types of colorimetric tests are commonly used:

(1) The pH is a measure of the concentration of hydrogen ions (the acidity of a solution) determined by the reaction of an indicator that varies in color, depending on hydrogen ion levels in the water.

(2) Tests which determine a concentration of an element or compound are based on Beer's Law. Simply, this law states that the higher the concentration of a substance, the darker the color produced in the test reaction, and therefore the more light absorbed. Assuming a constant viewpath, the absorption increases exponentially with concentration.

10.6.3 VISUAL METHODS

The Octet Comparator uses standards that are mounted in a plastic comparator block. It employs eight permanent translucent color standards and built-in filters to eliminate optical distortion. The sample is compared using either of two viewing windows. Two devices that can be used with the comparator are the B-color Reader, which neutralizes color or turbidity in water samples, and viewpath, which intensifies faint colors of low concentrations for easy distinction.

10.6.4 ELECTRONIC METHODS

Although the human eye is capable of differentiating color intensity, interpretation is quite subjective. Electronic colorimeters consist of a light source that passes through a sample and is measured on a photodetector with an analog or digital readout.

10.6.5 ELECTRONIC METERS

Besides electronic colorimeters, specific electronic instruments are manufactured for lab and field determination of many water quality factors, including pH, total dissolved solids (TDS)/conductivity, dissolved oxygen, temperature, and turbidity.

10.7 DISSOLVED OXYGEN (DO) AND BIOCHEMICAL OXYGEN DEMAND (BOD)[15]

A stream system (in this case, the hypothetical one that provides the source

---

[15]*Note:* In this section and the sections that follow, we discuss several water quality factors that are routinely monitored in drinking water operations. We do not discuss the actual test procedures to analyze each water quality factor; we refer you to the latest edition of *Standard Methods* for the correct procedure to use in conducting these tests.

of water used in our discussion) produces and consumes oxygen. It gains oxygen from the atmosphere and from plants as a result of photosynthesis. Because of running water's churning, it dissolves more oxygen than does still water, such as in a reservoir behind a dam. Respiration by aquatic animals, decomposition, and various chemical reactions consume oxygen.

Oxygen is actually poorly soluble in water. Its solubility is related to pressure and temperature. In water supply systems, *dissolved oxygen* (DO) in raw water is considered the necessary element to support life of many aquatic organisms. From the drinking water practitioner's point of view, DO is an important indicator of the water treatment process, and an important factor in corrosivity.

Wastewater from sewage treatment plants often contains organic materials that are decomposed by microorganisms, which use oxygen in the process. (The amount of oxygen consumed by these organisms in breaking down the waste is known as the *biochemical oxygen demand* (BOD). We include a discussion of BOD and how to monitor it at the end of this section.) Other sources of oxygen-consuming waste include stormwater runoff from farmland or urban streets, feedlots, and failing septic systems.

Oxygen is measured in its dissolved form as dissolved oxygen (DO). If more oxygen is consumed than produced, dissolved oxygen levels decline and some sensitive animals may move away, weaken, or die.

DO levels fluctuate over a 24-hour period and seasonally. They vary with water temperature and altitude. Cold water holds more oxygen than warm water (Table 10.5), and water holds less oxygen at higher altitudes. Thermal discharges (such as water used to cool machinery in a manufacturing plant or a power plant) raise the temperature of water and lower its oxygen content. Aquatic animals are most vulnerable to lowered DO levels in the early morning on hot summer days when stream flows are low, water temperatures are high, and aquatic plants have not been producing oxygen since sunset.

10.7.1 SAMPLING AND EQUIPMENT-CONSIDERATIONS

In contrast to lakes, where DO levels are most likely to vary vertically in the water column, changes in DO in rivers and streams move horizontally along the course of the waterway. This is especially true in smaller, shallow streams. In larger, deeper rivers, some vertical stratification of dissolved oxygen might occur. The DO levels in and below riffle areas, waterfalls, or dam spillways are typically higher than those in pools and slower-moving stretches. If you wanted to measure the effect of a dam, sampling for DO behind the dam, immediately below the spillway, and upstream of the dam would be important. Since DO levels are critical to fish, a good place to sample is in the pools that fish tend to favor, or in the spawning areas they use.

An hourly time profile of DO levels at a sampling site is a valuable set of

TABLE 10.5. Maximum Dissolved Oxygen Concentrations versus Temperature Variations.

Temperature °C	DO (mg/L)	Temperature °C	DO (mg/L)
0	14.60	23	8.56
1	14.19	24	8.40
2	13.81	25	8.24
3	13.44	26	8.09
4	13.09	27	7.95
5	12.75	28	7.81
6	12.43	29	7.67
7	12.12	30	7.54
8	11.83	31	7.41
9	11.55	32	7.28
10	11.27	33	7.16
11	11.01	34	7.05
12	10.76	35	6.93
13	10.52	36	6.82
14	10.29	37	6.71
15	10.07	38	6.61
16	9.85	39	6.51
17	9.65	40	6.41
18	9.45	41	6.31
19	9.26	42	6.22
20	9.07	43	6.13
21	8.90	44	6.04
22	8.72	45	5.95

data, because it shows the change in DO levels from the low point (just before sunrise) to the high point (sometime near midday). However, this might not be practical for a volunteer monitoring program. Note the time of your DO sampling to help judge when in the daily cycle the data were collected.

DO is measured either in milligrams per liter (mg/L) or "percent saturation." Milligrams per liter is the amount of oxygen in a liter of water. Percent saturation is the amount of oxygen in a liter of water relative to the total amount of oxygen that the water can hold at that temperature.

DO samples are collected using a special BOD bottle: a glass bottle with a "turtleneck" and a ground glass stopper. You can fill the bottle directly in the stream if the stream is wadeable or boatable, or you can use a sampler dropped from a bridge or boat into water deep enough to submerse it. Samplers can be made or purchased.

Dissolved oxygen is measured primarily either by using some variation of the Winkler method, or by using a meter and probe.

10.7.1.1 Winkler Method

The *Winkler method* involves filling a sample bottle completely with water (no air is left to bias the test). The dissolved oxygen is then "fixed" using a series of reagents that form a titrated acid compound. Titration involves the drop-by-drop addition of a reagent that neutralizes the acid compound, causing a change in the color of the solution. The point at which the color changes is the "endpoint" and is equivalent to the amount of oxygen dissolved in the sample. The sample is usually fixed and titrated in the field at the sample site. Preparing the sample in the field and delivering it to a lab for titration is possible.

Dissolved oxygen field kits using the Winkler method are relatively inexpensive, especially compared to a meter and probe. Field kits run between $35 and $200, and each kit comes with enough reagents to run 50 to 100 DO tests. Replacement reagents are inexpensive, and you can buy them already measured out for each test in plastic pillows.

You can also buy the reagents in larger quantities in bottles, and measure them out with a volumetric scoop. The pillows' advantage is that they have a longer shelf life and are much less prone to contamination or spillage. Buying larger quantities in bottles has the advantage of considerably lower cost per test.

The major factor in the expense of the kits is the method of titration used—eyedropper, syringe-type titrator, or digital titrator. Eyedropper and syringe-type titration is less precise than digital titration, because a larger drop of titrant is allowed to pass through the dropper opening, and on a micro-scale, the drop size (and thus the volume of titrant) can vary from drop to drop. A digital titrator or a burette (a long glass tube with a tapered tip like a pipette) permits much more precision and uniformity in the amount of titrant it allows to pass.

If a high degree of accuracy and precision in DO results is required, a digital titrator should be used. A kit that uses an eyedropper-type or syringe-type titrator is suitable for most other purposes. The lower cost of this type of DO field kit might be attractive if several teams of samplers and testers at multiple sites at the same time are relied on.

10.7.1.2 Meter and Probe

A *dissolved oxygen meter* is an electronic device that converts signals from a probe placed in the water into units of DO in milligrams per liter. Most meters and probes also measure temperature. The probe is filled with a salt solution and has a selectively permeable membrane that allows DO to pass from the stream water into the salt solution. The DO that has diffused into the salt solution changes the electric potential of the salt solution, and this change is sent by electric cable to the meter, which converts the signal to milligrams per liter on a scale that the volunteer can read.

DO meters are expensive compared to field kits that use the titration method. Meter/probe combinations run between $500 and $1200, including a long cable to connect the probe to the meter. The advantage of a meter/probe is that DO and temperature can be quickly read at any point where the probe is inserted into the stream. DO levels can be measured at a certain point on a continuous basis. The results are read directly as milligrams per liter, unlike the titration methods, in which the final titration result might have to be converted by an equation to milligrams per liter.

However, DO meters are more fragile than field kits, and repairs to a damaged meter can be costly. The meter/probe must be carefully maintained, and must be calibrated before each sample run, and if many tests are done, between sampling. Because of the expense, a small waterworks might only have one meter/probe, which means that only one team of samplers can sample DO, and they must test all the sites. With field kits, on the other hand, several teams can sample simultaneously.

10.7.1.3 Laboratory Testing of Dissolved Oxygen

If a meter and probe are used, the testing must be done in the field because dissolved oxygen levels in a sample bottle change quickly from the decomposition of organic material by microorganisms, or the production of oxygen by algae and other plants in the sample. This lowers the DO reading. If a variation of the Winkler method is used, "fixing" the sample in the field then delivering it to a lab for titration is possible. This might be preferable if sampling is conducted under adverse conditions or if time spent collecting samples is an issue. Titrating samples in the lab is a little easier, and more quality control is possible because the same person can do all the titrations.

10.7.2 WHAT IS BIOCHEMICAL OXYGEN DEMAND AND WHY IS IT IMPORTANT?

Biochemical oxygen demand (BOD) measures the amount of oxygen consumed by microorganisms in decomposing organic matter in stream water. BOD also measures the chemical oxidation of inorganic matter (the extraction of oxygen from water via chemical reaction). A test is used to measure the amount of oxygen consumed by these organisms during a specified period of time (usually five days at 20°C). The rate of oxygen consumption in a stream is affected by a number of variables: temperature, pH, the presence of certain kinds of microorganisms, and the type of organic and inorganic material in the water.

BOD directly affects the amount of dissolved oxygen in rivers and streams. The greater the BOD, the more rapidly oxygen is depleted in the stream, leaving less oxygen available to higher forms of aquatic life. The consequences of high BOD are the same as those for low dissolved oxygen: aquatic organisms become stressed, suffocate, and die. Most river waters used as water supplies have a BOD of less than 7 mg/L; therefore dilution is not necessary.

Sources of BOD include leaves and woody debris; dead plants and animals; animal manure; effluents from pulp and paper mills, wastewater treatment plants, feedlots, and food-processing plants; failing septic systems; and urban stormwater runoff.

Note: To evaluate raw water's potential for use as a drinking water supply, it is usually sampled, analyzed, and tested for biochemical oxygen demand when turbid, polluted water is the only source available.

10.7.2.1 Sampling Considerations

BOD is affected by the same factors that affect dissolved oxygen (see above). Aeration of stream water—by rapids and waterfalls, for example—will accelerate the decomposition of organic and inorganic material. Therefore, BOD levels at a sampling site with slower, deeper waters might be higher for a given column of organic and inorganic material than the levels for a similar site in high aerated waters.

Chlorine can also affect BOD measurement by inhibiting or killing the microorganisms that decompose the organic and inorganic matter in a sample. If sampling in chlorinated waters (such as those below the effluent from a sewage treatment plant), neutralizing the chlorine with sodium thiosulfate is necessary (see *Standard Methods*).

BOD measurement requires taking two samples at each site. One is tested immediately for dissolved oxygen, and the second is incubated in the dark at 20°C for five days, then tested for the amount of dissolved oxygen remaining. The difference in oxygen levels between the first test and the second test [in milligrams per liter (mg/L)] is the amount of BOD. This represents the amount of oxygen consumed by microorganisms and used to break down the organic matter present in the sample bottle during the incubation period. Because of the five-day incubation, the tests are conducted in a laboratory.

Sometimes by the end of the five-day incubation period, the dissolved oxygen level is zero. This is especially true for rivers and streams with a lot of organic pollution. Since knowing when the zero point was reached is not possible, determining the BOD level is also impossible. In this case, diluting the original sample by a factor that results in a final dissolved oxygen level of at least 2 mg/L is necessary. Special dilution water should be used for the dilutions (see *Standard Methods*).

Some experimentation is needed to determine the appropriate dilution factor for a particular sampling site. The final result is the difference in dissolved oxygen between the first measurement and the second, after multiplying the second result by the dilution factor. *Standard Methods* prescribes all phases of procedures and calculations for BOD determination. A BOD test is not required for monitoring water supplies.

10.7.3 TEMPERATURE

We stated earlier (see Section 7.4) that an ideal water supply should have, at all times, an almost constant temperature or one with minimum variation. Knowing the temperature of the water supply is important because the rates of biological and chemical processes depend on it.

Temperature affects the oxygen content of the water (oxygen levels become lower as temperature increases); the rate of photosynthesis by aquatic plants; the metabolic rates of aquatic organisms; and the sensitivity of organisms to toxic wastes, parasites, and diseases.

Causes of temperature change include weather, removal of shading streambank vegetation, impoundments (a body of water confined by a barrier, such as a dam), discharge of cooling water, urban stormwater, and groundwater inflows to the stream.

10.7.3.1 Sampling and Equipment Considerations

Temperature in a stream varies with width and depth, and the temperature of well-sunned portions of a stream can be significantly higher than the shaded portion of the water on a sunny day. In a small stream, the temperature will be relatively constant as long as the stream is uniformly in sun or shade. In a large stream, temperature can vary considerably with width and depth, regardless of shade. If safe to do so, temperature measurements should be collected at varying depths and across the surface of the stream to obtain vertical and horizontal temperature profiles. This can be done at each site at least once to determine the necessity of collecting a profile during each sampling visit. Temperature should be measured at the same place every time.

Temperature is measured in the stream with a thermometer or a meter. Alcohol-filled thermometers are preferred over mercury-filled because they are less hazardous if broken. Armored thermometers for field use can withstand more abuse than unprotected glass thermometers and are worth the additional expense. Meters for other tests [such as pH (acidity) or dissolved oxygen] also measure temperature and can be used instead of a thermometer.

10.7.4 HARDNESS

As we pointed out in Section 7.9, *hardness* refers primarily to the amount of calcium and magnesium in the water. Calcium and magnesium enter water mainly by leaching of rocks. Calcium is an important component of aquatic plant cell walls and the shells and bones of many aquatic organisms. Magnesium is an essential nutrient for plants and is a component of the chlorophyll molecule.

Hardness test kits express test results in ppm of $CaCO_3$, but these results can be converted directly to calcium or magnesium concentrations:

$$\text{Calcium Hardness as ppm } CaCO_3 \times 0.40 = \text{ppm Ca} \qquad (10.1)$$

$$\text{Magnesium Hardness as ppm } CaCO_3 \times 0.24 = \text{ppm Mg} \qquad (10.2)$$

Note: Because of less contact with soil minerals and more contact with rain, surface raw water is usually softer than groundwater.

As a general rule of thumb, when hardness is greater than 150 mg/L, softening treatment may be required for public water systems. Hardness determination via testing is required to ensure efficiency of treatment.

10.7.4.1 Measuring Hardness

A hardness test follows a procedure similar to an alkalinity test (see Section 10.7.11), but the reactions involved are different. The sample must be carefully measured, then a buffer is added to the sample to correct pH for the test and an indicator to signal the titration end point.

The indicator reagent is normally blue in a sample of pure water, but if calcium or magnesium ions are present in the sample, the indicator combines with them to form a red-colored complex. The titrant in this test is EDTA (ethylenediaminetetraacetic acid, used with its salts in the titration method), a "chelant" which actually "pulls" the calcium and magnesium ions away from the red-colored complex. The EDTA is added dropwise to the sample until all the calcium and magnesium ions have been "chelated" away from the complex and the indicator returns to its normal blue color. The amount of EDTA required to cause the color change is a direct indication of the amount of calcium and magnesium ions in the sample.

Some hardness kits include an additional indicator that is specific for calcium. This type of kit will provide three readings: total hardness, calcium hardness, and magnesium hardness. For interference, precision, and accuracy, consult the latest edition of *Standard Methods.*

10.7.5 pH

We pointed out in Section 7.7 that *pH* is a term used to indicate the alkalinity or acidity of a substance as ranked on a scale from 1.0 to 14.0. Acidity increases as the pH gets lower. Figure 10.3 present the pH of some common liquids.

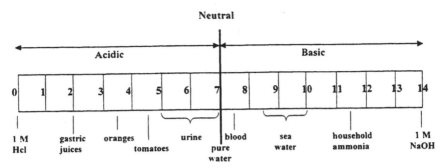

Figure 10.3 pH of selected liquids.

10.7.5.1 Analytical and Equipment Considerations

The pH can be analyzed in the field or in the lab. If analyzed in the lab, it must be measured within two hours of the sample collection, because the pH will change from the carbon dioxide from the air as it dissolves in the water, bringing the pH toward 7.

If your program requires a high degree of accuracy and precision in pH results, the pH should be measured with a laboratory quality pH meter and electrode. Meters of this quality range in cost from around $250 to $1000. Color comparators and pH "pocket pals" are suitable for most other purposes. The cost of either of these is in the $50 range. The lower cost of the alternatives might be attractive if multiple samplers are used to sample several sites at the same time.

10.7.5.1.1 pH Meters

A pH meter measures the electric potential (millivolts) across an electrode when immersed in water. This electric potential is a function of the hydrogen ion activity in the sample; therefore, pH meters can display results in either millivolts (mV) or pH units.

A pH meter consists of a *potentiometer*, which measures electric potential where it meets the water sample; a reference electrode, which provides a constant electric potential; and a *temperature compensating device*, which adjusts the readings according to the temperature of the sample (since pH varies with temperature). The reference and glass electrodes are frequently combined into a single probe called a *combination electrode*.

A wide variety of meters are available, but the most important part of the pH meter is the electrode. Thus, purchasing a good, reliable electrode and following the manufacturer's instructions for proper maintenance is important. Infrequently used or improperly maintained electrodes are subject to corrosion, which makes them highly inaccurate.

10.7.5.1.2 pH "Pocket Pals" and Color Comparators

pH "pocket pals" are electronic handheld "pens" that are dipped in the water, providing a digital readout of the pH. They can be calibrated to only one pH buffer. (Lab meters, on the other hand, can be calibrated to two or more buffer solutions and thus are more accurate over a wide range of pH measurements.)

Color comparators involve adding a reagent to the sample that colors the sample water. The intensity of the color is proportional to the pH of the sample, then matched against a standard color chart. The color chart equates particular colors to associated pH values, which can be determined by matching the colors from the chart to the color of the sample.

For instructions on how to collect and analyze samples, refer to *Standard Methods*.

10.7.6 TURBIDITY

We discussed *turbidity* in Section 7.5. Turbidity is a measure of water clarity—how much the material suspended in water decreases the passage of light through the water. Suspended materials include soil particles (clay, silt, and sand), algae, plankton, microbes, and other substances. These materials are typically in the size range of 0.004 mm (clay) to 1.0 mm (sand). Turbidity can affect the color of the water.

Higher turbidity increases water temperatures, because suspended particles absorb more heat. This in turn reduces the concentration of dissolved oxygen (DO) because warm water holds less DO than cold. Higher turbidity also reduces the amount of light penetrating the water, which reduces photosynthesis and the production of DO. Suspended materials can clog fish gills, reducing resistance to disease in fish, lowering growth rates, and affecting egg and larval development. As the particles settle, they can blanket the stream bottom (especially in slower waters) and smother fish eggs and benthic macroinvertebrates. Sources of turbidity include:

- soil erosion
- waste discharge
- urban runoff
- eroding stream banks
- large numbers of bottom feeders (such as carp), which stir up bottom sediments
- excessive algal growth

10.7.6.1 Sampling and Equipment Considerations

Turbidity can be useful as an indicator of the effects of runoff from construction, agricultural practices, logging activity, discharges, and other sources. Turbidity often increases sharply during a rainfall, especially in developed watersheds, which typically have relatively high proportions of impervious surfaces. The flow of stormwater runoff from impervious surfaces rapidly increases stream velocity, which increases the erosion rates of streambanks and channels. Turbidity can also rise sharply during dry weather if Earth-disturbing activities occur in or near a stream without erosion control practices in place.

Regular monitoring of turbidity can help detect trends that might indicate increasing erosion in developing watersheds. However, turbidity is closely related to stream flow and velocity and should be correlated with these factors. Comparisons of the change in turbidity over time, therefore, should be made at the same point at the same flow.

Turbidity is not a measurement of the amount of suspended solids present or the rate of sedimentation of a stream since it measures only the amount of light that is scattered by suspended particles. Measurement of total solids is a more direct measurement of the amount of material suspended and dissolved in water.

Turbidity is generally measured by using a turbidity meter. Volunteer programs may also take samples to a lab for analysis. Another approach is to measure transparency (an integrated measure of light scattering and absorption) instead of turbidity. Water clarity/transparency can be measured using a *Secchi disk* (see Figure 10.4) or transparency tube. The Secchi disk can only be used in deep, slow moving rivers (see Section 10.7.6.1.1); the transparency tube (a comparatively new development) is gaining acceptance in and around the country, but is not yet in wide use (see Section 10.7.6.1.2).

A turbidity meter consists of a light source that illuminates a water sample, and a photoelectric cell that measures the intensity of light scattered at a 90° angle by the particles in the sample. It measures turbidity in nephelometric

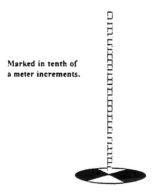

Marked in tenth of
a meter increments.

Figure 10.4 Using a Secchi disk to measure transparency. Lower the disk into the water until it is no longer visible. That point is the Secchi disk depth.

turbidity units (NTUs). Meters can measure turbidity over a wide range—from 0 to 1000 NTUs. A clear mountain stream might have a turbidity of around 1 NTU, whereas a large river like the Mississippi might have a dry-weather turbidity of 10 NTUs. Because these values can jump into hundreds of NTUs during runoff events, the turbidity meter to be used should be reliable over the range in which you will be working. Meters of this quality cost about $800. Many meters in this price range are designed for field or lab use.

Although turbidity meters can be used in the field, samplers might want to collect samples and take them to a central point for turbidity measurements, because of the expense of the meter. Most programs can afford only one and would have to pass it along from site to site, complicating logistics and increasing the risk of damage to the meter. Meters also include glass cells that must remain optically clear and free of scratches for operation.

Samplers can also take turbidity samples to a lab for meter analysis at a reasonable cost.

10.7.6.1.1 Using a Secchi Disk

A Secchi disk is a black and white disk that is lowered by hand into the water to the depth at which it vanishes from sight (see Figure 10.4). The distance to vanishing is then recorded—the clearer the water, the greater the distance. Secchi disks are simple to use and inexpensive. For river monitoring they have limited use, however, because in most cases the river bottom will be visible and the disk will not reach a vanishing point. Deeper, slower moving rivers are the most appropriate places for Secchi disk measurement, although the current might require that the disk be extra-weighted so it does not sway and make measurement difficult. Secchi disks cost about $50, but can be homemade.

The line attached to the Secchi disk must be marked in waterproof ink according to units designated by the sampling program. Many programs require samplers to measure to the nearest 1/10 meter. Meter intervals can be tagged (e.g., with duct tape) for ease of use.

To measure water clarity with a Secchi disk:

- Check to make sure that the Secchi disk is securely attached to the measured line.
- Lean over the side of the boat and lower the Secchi disk into the water, keeping your back to the sun to block glare.
- Lower the disk until it disappears from view. Lower it one third of a meter and then slowly raise the disk until it just reappears. Move the disk up and down until you find the exact vanishing point.
- Attach a clothespin to the line at the point where the line enters the

water. Record the measurement on your data sheet. Repeating the measurement provides you with a quality control check.

The key to consistent results is to train samplers to follow standard sampling procedures, and if possible, have the same individual take the reading at the same site throughout the season.

10.7.6.1.2 Transparency Tube

Pioneered by Australia's Department of Conservation, the *transparency tube* is a clear, narrow plastic tube marked in units with a dark pattern painted on the bottom. Water is poured into the tube until the pattern disappears. Some U.S. volunteer monitoring programs [e.g., the Tennessee Valley Authority (TVA) Clean Water Initiative and the Minnesota Pollution Control Agency (MPCA)] are testing the transparency tube in streams and rivers. MPCA uses tubes marked in centimeters, and has found tube readings to relate fairly well to lab measurements of turbidity and total suspended solids, although it does not recommend the transparency tube for applications where precise and accurate measurement is required or in highly colored waters.

The TVA and MPCA recommended the following sampling considerations:

- Collect the sample in a bottle or bucket in mid-stream and at mid-depth if possible. Avoid stagnant water and sample as far from the shoreline as is safe. Avoid collecting sediment from the bottom of the stream.
- Face upstream as you fill the bottle or bucket.
- Take readings in open but shaded conditions. Avoid direct sunlight by turning your back to the sun.
- Carefully stir or swish the water in the bucket or bottle until it is homogeneous, taking care not to produce air bubbles (these scatter light and affect the measurement). Then pour the water slowly in the tube while looking down the tube. Measure the depth of the water column in the tube at the point where the symbol just disappears.

10.7.7 ORTHOPHOSPHATE

In Section 8.8, we discussed the nutrients phosphorus and nitrogen. Both phosphorus and nitrogen are essential nutrients for the plants and animals that make up the aquatic food web. Since phosphorus is the nutrient in short supply in most freshwater systems, even a modest increase in phosphorus can (under the right conditions) set off a whole chain of undesirable events in a stream, including accelerated plant growth, algae blooms, low dissolved oxygen, and the death of certain fish, invertebrates, and other aquatic animals.

Phosphorus comes from many sources, both natural and human. These include soil and rocks, wastewater treatment plants, runoff from fertilized lawns and cropland, failing septic systems, runoff from animal manure storage areas, disturbed land areas, drained wetlands, water treatment, and commercial cleaning preparations.

10.7.7.1 Forms of Phosphorus

Phosphorus has a complicated story. Pure, "elemental" phosphorus (P) is rare. In nature, phosphorus usually exists as part of a phosphate molecule (PO_4). Phosphorus in aquatic systems occurs as organic phosphate and inorganic phosphate. Organic phosphate consists of a phosphate molecule associated with a carbon-based molecule, as in plant or animal tissue. Phosphate that is not associated with organic material is inorganic, the form required by plants. Animals can use either organic or inorganic phosphate.

Both organic and inorganic phosphorus can either be dissolved in the water or suspended (attached to particles in the water column).

10.7.7.1.1 The Phosphorus Cycle

Phosphorus cycles though the environment, changing form as it does so (see Figure 10.5). Aquatic plants take in dissolved inorganic phosphorus as it becomes part of their tissues. Animals get the organic phosphorus they need by eating either aquatic plants, other animals, or decomposing plant and animal material.

In water bodies, as plants and animals excrete wastes or die, the organic phosphorus they contain sinks to the bottom, where bacterial decomposition converts it back to inorganic phosphorus, both dissolved and attached to

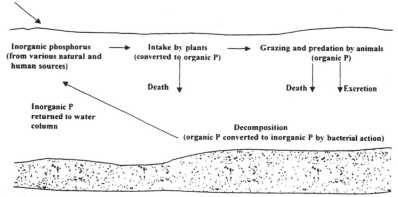

Figure 10.5 The phosphorus cycle. Phosphorus changes form as it cycles through the aquatic environment.

particles. This inorganic phosphorus gets back into the water column when the bottom is stirred up by animals, human activity, chemical interactions, or water currents. Then it is taken up by plants and the cycle begins again.

In a stream system, the phosphorus cycle tends to move phosphorus downstream as the current carries decomposing plant and animal tissue and dissolved phosphorus. It becomes stationary only when it is taken up by plants or is bound to particles that settle to the bottom of ponds.

In the field of water quality chemistry, phosphorus is described by several terms. Some of these terms are chemistry based (referring to chemically based compounds), and others are methods based (they describe what is measured by a particular method).

The term *orthophosphate* is a chemistry-based term that refers to the phosphate molecule all by itself. *Reactive phosphorus* is a corresponding method-based term that describes what is actually being measured when the test for orthophosphate is being performed. Because the lab procedure isn't quite perfect, mostly orthophosphate is obtained along with a small fraction of some other forms.

More complex inorganic phosphate compounds are referred to as *condensed phosphates* or *polyphosphates*. The method-based term for these forms is *acid hydrolyzable*.

10.7.7.1.2 Monitoring Phosphorus

Monitoring phosphorus is challenging because it involves measuring very low concentrations, down to 0.01 milligrams per liter (mg/L) or even lower. Even such very low concentrations of phosphorus can have a dramatic impact on streams. Less sensitive methods should be used only to identify serious problem areas.

While many tests for phosphorus exist, only four are likely to be performed by most monitors.

(1) The *total orthophosphate* test is largely a measure of orthophosphate. Because the sample is not filtered, the procedure measures both dissolved and suspended orthophosphate. The USEPA-approved method for measuring total orthophosphate is known as the ascorbic acid method. Briefly, a reagent (either liquid or powder) containing ascorbic acid and ammonium molybdate reacts with orthophosphate in the sample to form a blue compound. The intensity of the blue color is directly proportional to the amount of orthophosphate in the water.

(2) The *total phosphorus* test measures all the forms of phosphorus in the sample (orthophosphate, condensed phosphate, and organic phosphate) by first "digesting" (heating and acidifying) the sample to convert all the other forms to orthophosphate, then the orthophosphate is measured by

the ascorbic acid method. Because the sample is not filtered, the procedure measures both dissolved and suspended orthophosphate.

(3) The *dissolved phosphorus* test measures that fraction of the total phosphorus that is in solution in the water (as opposed to being attached to suspended particles). It is determined by first filtering the sample, then analyzing the filtered sample for total phosphorus.

(4) *Insoluble phosphorus* is calculated by subtracting the dissolved phosphorus result from the total phosphorus result.

All these tests have one thing in common—they all depend on measuring orthophosphate. The total orthophosphate test measures the orthophosphate that is already present in the sample. The others measure that which is already present and that which is formed when the other forms of phosphorus are converted to orthophosphate by digestion.

10.7.7.1.3 Sampling and Equipment Considerations

Monitoring phosphorus involves two basic steps:

- collecting a water sample
- analyzing it in the field or lab for one of the types of phosphorus described above

This text does not address laboratory methods. Refer to *Standard Methods*.

10.7.7.1.3.1 SAMPLE CONTAINERS

Sample containers made of either some form of plastic or Pyrex® glass are acceptable to the USEPA. Because phosphorus molecules have a tendency to "adsorb" (attach) to the inside surface of sample containers, if containers are to be reused, they must be acid-washed to remove adsorbed phosphorus. The container must be able to withstand repeated contact with hydrochloric acid. Plastic containers, either high-density polyethylene or polypropylene, might be preferable to glass from a practical standpoint because they are better able to withstand breakage. Some programs use disposable, sterile, plastic Whirl-pak® bags. The size of the container depends on the sample amount needed for the phosphorus analysis method chosen, and the amount needed for other analyses to be performed.

10.7.7.1.3.2 DEDICATED LABWARE

All containers that will hold water samples or come into contact with reagents used in this test must be dedicated. They should not be used for other tests, to eliminate the possibility that reagents containing phosphorus will contaminate the labware. All labware should be acid-washed.

The only form of phosphorus this text recommends for field analysis is total orthophosphate, which uses the ascorbic acid method on an untreated sample. Analysis of any of the other forms requires adding potentially hazardous reagents, heating the sample to boiling, and using too much time and too much equipment to be practical. In addition, analysis for other forms of phosphorus is prone to errors and inaccuracies in field situations. Pretreatment and analysis for these other forms should be handled in a laboratory.

10.7.7.1.3.3 ASCORBIC ACID METHOD

In the ascorbic acid method, a combined liquid or prepackaged powder reagent consisting of sulfuric acid, potassium antimonyl tartrate, ammonium molybdate, and ascorbic acid (or comparable compounds) is added to either 50 or 25 mL of the water sample. This colors the sample blue in direct proportion to the amount of orthophosphate in the sample. Absorbence or transmittance is then measured after 10 minutes, but before 30 minutes, using a color comparator with a scale in milligrams per liter that increases with the increase in color hue, or an electronic meter that measures the amount of light absorbed or transmitted at a wavelength of 700–880 nanometers (again, depending on manufacturer's directions).

A color comparator may be useful for identifying heavily polluted sites with high concentrations (greater than 0.1 mg/L). However, matching the color of a treated sample to a comparator can be very subjective, especially at low concentrations, and can lead to variable results.

A field spectrophotometer or colorimeter with a 2.5-cm light path and an infrared photocell (set for a wavelength of 700–880 nm) is recommended for accurate determination of low concentrations (between 0.2 and 0.02 mg/L). Use of a meter requires that a prepared known standard concentration be analyzed ahead of time to convert the absorbence readings of a stream sample to milligrams per liter, or that the meter reads directly in milligrams per liter.

For information on how to prepare standard concentrations and on how to collect and analyze samples, refer to *Standard Methods* and USEPA's *Methods for Chemical Analysis of Water and Wastes* (2nd ed., 1991, Method 365.2).

10.7.8 NITRATES

As we discussed in Section 8.8, *nitrates* are a form of nitrogen found in several different forms in terrestrial and aquatic ecosystems. These forms of nitrogen include ammonia (NH_3), nitrates (NO_3), and nitrites (NO_2). Nitrates are essential plant nutrients, but excess amounts can cause significant water quality problems. Together with phosphorus, nitrates in excess amounts can accelerate eutrophication, causing dramatic increases in aquatic plant growth and changes in the types of plants and animals that live in the stream. This, in turn, affects dissolved oxygen, temperature, and other indicators. Excess nitrates can cause hypoxia (low levels of dissolved oxygen) and can become toxic to warm-blooded animals at higher concentrations (10 mg/L or higher) under certain conditions. The natural level of ammonia or nitrate in surface water is typically low (less than 1 mg/L); in the effluent of wastewater treatment plants, it can range up to 30 mg/L.

Sources of nitrates include wastewater treatment plants, runoff from fertilized lawns and cropland, failing on-site septic systems, runoff from animal manure storage areas, and industrial discharges that contain corrosion inhibitors.

10.7.8.1 Sampling and Equipment Considerations

Nitrates from land sources end up in rivers and streams more quickly than other nutrients like phosphorus, because they dissolve in water more readily than phosphorus, which has an attraction for soil particles. As a result, nitrates serve as a better indicator of the possibility of a source of sewage or manure pollution during dry weather.

Water that is polluted with nitrogen-rich organic matter might show low nitrates. Decomposition of the organic matter lowers the dissolved oxygen level, which in turn slows the rate at which ammonia is oxidized to nitrite (NO_2) and then to nitrate (NO_3). Under such circumstances, monitoring for nitrites or ammonia (considerably more toxic to aquatic life than nitrate) might be also necessary. (See *Standard Methods* sections 4500-NH_3 and 4500-NH_2 for appropriate nitrite methods.)

Water samples to be tested for nitrate should be collected in glass or polyethylene containers that have been prepared by using Method B (see Section 10.5.1.2).

Most monitoring programs usually use two methods for nitrate testing: the cadmium reduction method and the nitrate electrode. The more commonly used cadmium reduction method produces a color reaction measured either by comparison to a color wheel or by use of a spectrophotometer. A few programs also use a nitrate electrode, which can measure in the range of 0 to 100 mg/L nitrate. A newer colorimetric immunoassay technique for nitrate screening is also now available.

10.7.8.1.1 Cadmium Reduction Method

The *cadmium reduction method* is a colorimetric method that involves contact of the nitrate in the sample with cadmium particles, which cause nitrates to be converted to nitrites. The nitrites then react with another reagent to form a red color, in proportional intensity to the original amount of nitrate. The color is measured either by comparison to a color wheel with a scale in milligrams per liter that increases with the increase in color hue, or by use of an electronic spectrophotometer that measures the amount of light absorbed by the treated sample at a 543-nanometer wavelength. The absorbence value converts to the equivalent concentration of nitrate against a standard curve. Methods for making standard solutions and standard curves are presented in *Standard Methods*.

This curve should be created by the program advisor before each sampling run. The curve is developed by making a set of standard concentrations of nitrate, reacting them, and developing the corresponding color, then plotting the absorbence value for each concentration against concentration. A standard curve could also be generated for the color wheel. Use of the color wheel is appropriate only if nitrate concentrations are greater than 1 mg/L. For concentrations below 1 mg/L, use a spectrophotometer. Matching the color of a treated sample at low concentrations to a color wheel (or cubes) can be very subjective and can lead to variable results. Color comparators can, however, be effectively used to identify sites with high nitrates.

This method requires that the samples being treated are clear. If a sample is turbid, filter it through a 0.45-micron filter. Be sure to test to make sure the filter is nitrate-free. If copper, iron, or other metals are present in concentrations above several mg/L, the reaction with the cadmium will slow down and the reaction time must be increased.

The reagents used for this method are often prepackaged for different ranges, depending on the expected concentration of nitrate in the stream. For example, the Hach Company provides reagents for the following ranges: low (0 to 0.40 mg/L), medium (0 to 15 mg/L), and high (0 to 30 mg/L). Determining the appropriate range for the stream being monitored is important.

10.7.8.1.2 Nitrate Electrode Method

A nitrate electrode (used with a meter) is similar in function to a dissolved oxygen meter. It consists of a probe with a sensor that measures nitrate activity in the water; this activity affects the electric potential of a solution in the probe. This change is then transmitted to the meter, which converts the electric signal to a scale that is read in millivolts; then the millivolts are converted to mg/L of nitrate by plotting them against a standard curve. The accuracy of the electrode can be affected by high concentrations of chloride or bicarbonate ions in the sample water. Fluctuating pH levels can also affect the meter reading.

Nitrate electrodes and meters are expensive compared to field kits that employ the cadmium reduction method. (The expense is comparable, however, if a spectrophotometer is used rather than a color wheel.) Meter/probe combinations run between $700 and $1200, including a long cable to connect the probe to the meter. If the program has a pH meter that displays readings in millivolts, it can be used with a nitrate probe and no separate nitrate meter is needed. Results are read directly as milligrams per liter.

Although nitrate electrodes and spectrophotometers can be used in the field, they have certain disadvantages. These devices are more fragile than the color comparators and are therefore more at risk of breaking in the field. They must be carefully maintained and must be calibrated before each sample run, and if many tests are being run, between samplings. This means that samples are best tested in the lab. Note that samples to be tested with a nitrate electrode should be at room temperature, whereas color comparators can be used in the field with samples at any temperature.

10.7.9 TOTAL SOLIDS

Total solids (see Section 7.6) are dissolved solids plus suspended and settleable solids in water. In stream water, dissolved solids consist of calcium, chlorides, nitrate, phosphorus, iron, sulfur, and other ions—particles that will pass through a filter with pores of around 2 microns (0.002 cm) in size. Suspended solids include silt and clay particles, plankton, algae, fine organic debris, and other particulate matter. These are particles that will not pass through a 2-micron filter.

The concentration of total dissolved solids affects the water balance in the cells of aquatic organisms. An organism placed in water with a very low level of solids (distilled water, for example) swells because water tends to move into its cells, which have a higher concentration of solids. An organism placed in water with a high concentration of solids shrinks somewhat, because the water in its cells tends to move out. This in turn affects the organism's ability to maintain the proper cell density, making keeping its position in the water column difficult. It might float up or sink down to a depth to which it is not adapted, and it might not survive.

Higher concentrations of suspended solids can serve as carriers of toxics, which readily cling to suspended particles. This is particularly a concern where pesticides are being used on irrigated crops. Where solids are high, pesticide concentrations may increase well beyond those of the original application as the irrigation water travels down irrigation ditches. Higher levels of solids can also clog irrigation devices and might become so high that irrigated plant roots will lose water rather than gain it.

A high concentration of total solids will make drinking water unpalatable, and might have an adverse effect on people who are not used to drinking such water. Levels of total solids that are too high or too low can also reduce the efficiency of wastewater treatment plants, as well as the operation of industrial processes that use raw water.

Total solids affect water clarity. Higher solids decrease the passage of light through water, thereby slowing photosynthesis by aquatic plants. Water heats up more rapidly and holds more heat; this, in turn, might adversely affect aquatic life adapted to a lower temperature regime.

Sources of total solids include industrial discharges, sewage, fertilizers, road runoff, and soil erosion. Total solids are measured in milligrams per liter (mg/L).

10.7.9.1 Sampling and Equipment Considerations

Total solids are important to measure in areas where discharges from sewage treatment plants, industrial plants, or extensive crop irrigation may occur. In particular, streams and rivers in arid regions where water is scarce and evaporation is high tend to have higher concentrations of solids, and are more readily affected by human introduction of solids from land use activities.

Total solids measurements can be useful as an indicator of the effects of runoff from construction, agricultural practices, logging activities, sewage treatment plant discharges, and other sources. As with turbidity, concentrations often increase sharply during rainfall, especially in developed watersheds. They can also rise sharply during dry weather if earth-disturbing activities occur in or near the stream without erosion control practices in place. Regular monitoring of total solids can help detect trends that might indicate increasing erosion in developing watersheds. Total solids are closely related to stream flow and velocity, and should be correlated with these factors. Any change in total solids over time should be measured at the same site at the same flow.

Total solids are measured by weighing the amount of solids present in a known volume of sample; this is accomplished by weighing a beaker, filling it with a known volume, evaporating the water in an oven and completely drying the residue, then weighing the beaker with the residue. The total solids concentration is equal to the difference between the weight of the beaker with the residue and the weight of the beaker without it. Since the residue is so light in weight, the lab needs a balance that is sensitive to weights in the range of 0.0001 gram. Balances of this type are called analytical or Mettler balances, and they are expensive (around $3000). The technique requires that the beakers be kept in a desiccator, a sealed glass container that contains material that absorbs moisture and ensures that the weighing is not biased by water condensing on the beaker. Some desiccants change color to indicate moisture content.

The measurement of total solids cannot be done in the field. Samples must be collected using clean glass or plastic bottles, or Whirl-pak® bags and taken to a laboratory where the test can be run.

10.7.10 CONDUCTIVITY

Conductivity is a measure of the ability of water to pass an electrical current. Conductivity in water is affected by the presence of inorganic dissolved solids such as chloride, nitrate, sulfate, and phosphate anions (ions that carry a negative charge), or sodium, magnesium, calcium, iron, and aluminum cations (ions that carry a positive charge). Organic compounds like oil, phenol, alcohol, and sugar do not conduct electrical current very well, and therefore have a low conductivity when in water. Conductivity is also affected by temperature: the warmer the water, the higher the conductivity.

Conductivity in streams and rivers is affected primarily by the geology of the area through which the water flows. Streams that run through areas with granite bedrock tend to have lower conductivity because granite is composed of more inert materials that do not ionize (dissolve into ionic components) when washed into the water. On the other hand, streams that run through areas with clay soils tend to have higher conductivity, because of the presence of materials that ionize when washed into the water. Groundwater inflows can have the same effects, depending on the bedrock they flow through.

Discharges to streams can change the conductivity depending on their make-up. A failing sewage system would raise the conductivity because of the presence of chloride, phosphate, and nitrate; an oil spill would lower conductivity.

The basic unit of measurement of conductivity is the mho or siemens. Conductivity is measured in micromhos per centimeter (μmhos/cm) or microsiemens per centimeter (μs/cm). Distilled water has a conductivity in the range of 0.5 to 3 μmhos/cm. The conductivity of rivers in the United States generally ranges from 50 to 1500 μmhos/cm. Studies of inland freshwaters indicate that streams supporting good mixed fisheries have a range between 150 and 500 μmhos/cm. Conductivity outside this range could indicate that the water is not suitable for certain species of fish or macroinvertebrates. Industrial waters can range as high as 10,000 μmhos/cm.

10.7.10.1 Sampling and Equipment Considerations

Conductivity is useful as a general measure of stream water quality. Each stream tends to have a relatively constant range of conductivity that, once established, can be used as a baseline for comparison with regular conductivity measurements. Significant changes in conductivity could indicate that a discharge or some other source of pollution has entered a stream.

Conductivity is measured with a probe and a meter. Voltage is applied between two electrodes in a probe immersed in the sample water. The drop in voltage caused by the resistance of the water is used to calculate the conductivity per centimeter. The meter converts the probe measurement to micromhos per centimeter (μmhos/cm) and displays the result for the user.

Note: Some conductivity meters can also be used to test for total dissolved solids and salinity. The total dissolved solids concentration in milligrams per liter (mg/L) can also be calculated by multiplying the conductivity result by a factor between 0.55 and 0.9, which is empirically determined (see *Standard Methods* #2510).

Suitable conductivity meters cost about $350. Meters in this price range should also measure temperature and automatically compensate for temperature in the conductivity reading. Conductivity can be measured in the field or the lab. In most cases, collecting samples in the field and taking them to a lab for testing is probably better. In this way, several teams can collect samples simultaneously. If testing in the field is important, meters designed for field use can be obtained for around the same cost mentioned above.

If samples will be collected in the field for later measurement, the sample bottle should be a glass or polyethylene bottle that has been washed in phosphate-free detergent and rinsed thoroughly with both tap and distilled water. Factory-prepared Whirl-pak® bags may be used.

10.7.11 TOTAL ALKALINITY

In Section 7.8, we pointed out that *alkalinity* is a measure of the capacity of water to neutralize acids. Alkaline compounds in the water such as bicarbonates (baking soda is one type), carbonates, and hydroxides remove H^+ ions and lower the acidity of the water (which means increased pH). They usually do this by combining with the H^+ ions to make new compounds. Without this acid-neutralizing capacity, any acid added to a stream would cause an immediate change in the pH. Measuring alkalinity is important in determining a stream's ability to neutralize acidic pollution from rainfall or wastewater—one of the best measures of the sensitivity of the stream to acid inputs.

Alkalinity in streams is influenced by rocks and soils, salts, certain plant activities, and certain industrial wastewater discharges.

Total alkalinity is determined by measuring the amount of acid (e.g., sulfuric acid) needed to bring the sample to a pH of 4.2. At this pH all the alkaline compounds in the sample are "used up." The result is reported as milligrams per liter of calcium carbonate (mg/L $CaCO_3$).

10.7.11.1 Analytical and Equipment Considerations

For total alkalinity, a double end point titration using a pH meter (or pH "pocket pal") and a digital titrator or burette is recommended. This can be done in the field or in the lab. If alkalinity must be analyzed in the field, a digital titrator should be used instead of a burette, because burettes are fragile and more difficult to set up. The alkalinity method described below was developed by the Acid Rain Monitoring Project of the University of Massachusetts Water Resources Research Center.[16]

10.7.11.2 Burettes, Titrators, and Digital Titrators for Measuring Alkalinity

The total alkalinity analysis involves titration. In this test, titration is the addition of small, precise quantities of sulfuric acid (the reagent) to the sample, until the sample reaches a certain pH (known as an end point). The amount of acid used corresponds to the total alkalinity of the sample. Alkalinity can be measured using a burette, titrator, or digital titrator (described below).

- A *burette* is a long, graduated glass tube with a tapered tip like a pipette and a valve that opens to allow the reagent to drip out of the tube. The amount of reagent used is calculated by subtracting the original volume in the burette from the column left after the end point has been reached. Alkalinity is calculated based on the amount used.
- *Titrators* forcefully expel the reagent by using a manual or mechanical plunger. The amount of reagent used is calculated by subtracting the original volume in the titrator from the volume left after the end point has been reached. Alkalinity is then calculated based on the amount used or is read directly from the titrator.
- *Digital titrators* have counters that display numbers. A plunger is forced into a cartridge containing the reagent by turning a knob on the titrator. As the knob turns, the counter changes in proportion to the amount of reagent used. Alkalinity is then calculated based on the amount used. Digital titrators cost approximately $90.

Digital titrators and burettes allow for much more precision and uniformity in the amount of titrant that is used.

---

[16]From River Watch Network. *Total Alkalinity and pH Field and Laboratory Procedures.* Based on University of Massachusetts Acid Rain Monitoring Project, July 1, 1992.

10.7.12 FECAL BACTERIA[17]

As we discussed in Section 6.3.6, members of two bacteria groups (coliforms and fecal streptococci) are used as indicators of possible sewage contamination, because they are commonly found in human and animal feces. Although they are generally not harmful themselves, they indicate the possible presence of pathogenic (disease-causing) bacteria, viruses, and protozoans that also live in human and animal digestive systems. Their presence in streams suggests that pathogenic microorganisms might also be present, and that swimming in and/or eating shellfish from the waters might present a health risk. Since testing directly for the presence of a large variety of pathogens is difficult, time-consuming, and expensive, water is usually tested for coliforms and fecal streptococci instead. Sources of fecal contamination to surface waters include wastewater treatment plants, on-site septic systems, domestic and wild animal manure, and storm runoff.

In addition to the possible health risk associated with the presence of elevated levels of fecal bacteria, they can also cause cloudy water, unpleasant odors, and an increased oxygen demand.

10.7.12.1 Indicator Bacteria Types

The most commonly tested fecal bacteria indicators are total coliforms, fecal coliforms, *Escherichia coli*, fecal streptococci, and enterococci. All but *E. coli* are composed of a number of species of bacteria that share common characteristics, including shape, habitat, or behavior; *E. coli* is a single species in the fecal coliform group.

Total coliforms are widespread in nature. All members of the total coliform group can occur in human feces, but some can also be present in animal manure, soil, and submerged wood, and in other places outside the human body. The usefulness of total coliforms as an indicator of fecal contamination depends on the extent to which the bacteria species found are fecal and human in origin. For recreational waters, total coliforms are no longer recommended as an indicator. For drinking water, total coliforms are still the standard test, because their presence indicates contamination of a water supply by an outside source.

Fecal coliforms, a subset of total coliform bacteria, are more fecal-specific in origin. However, even this group contains a genus, *Klebsiella*, with species that are not necessarily fecal in origin. *Klebsiella* are commonly associated with textile and pulp and paper mill wastes. If these sources

[17]Much of the information in this section is from USEPA *Test Methods for* Escherichia coli *and Enterococci in Water by the Membrane Filter Procedure (Method #1103.1)*. EPA 600/4-85-076, 1985 and USEPA *Bacteriological Ambient Water Quality Criteria for Marine and Fresh Recreational Waters*. EPA 440/5-84-002. Cincinnati. OH: U.S. Environmental Protection Agency, Office of Research and Development, 1986.

discharge to a local stream, consideration should be given to monitoring more fecal and human-specific bacteria. For recreational waters, this group was the primary bacteria indicator until relatively recently, when the USEPA began recommending *E. coli* and enterococci as better indicators of health risk from water contact. Fecal coliforms are still being used in many states as indicator bacteria.

E. coli is a species of fecal coliform bacteria specific to fecal material from humans and other warm-blooded animals. The USEPA recommends *E. coli* as the best indicator of health risk from water contact in recreational waters; some states have changed their water quality standards and are monitoring accordingly.

Fecal streptococci generally occur in the digestive systems of humans and other warm-blooded animals. In the past, fecal streptococci were monitored together with fecal coliforms, and a ratio of fecal coliforms to streptococci was calculated. This ratio was used to determine whether the concentration was of human or nonhuman origin. However, this is no longer recommended as a reliable test.

Enterococci are a subgroup within the fecal streptococcus group. Enterococci are distinguished by their ability to survive in salt water, and in this respect they more closely mimic many pathogens than do the other indicators. Enterococci are typically more human-specific than the larger fecal streptococcus group. The USEPA recommends enterococci as the best indicator of health risk in salt water used for recreation, and as a useful indicator in freshwater as well.

10.7.12.2 Which Bacteria Should Be Monitored?

Which bacteria chosen for testing depends on what is to be determined. Does swimming in the local stream pose a health risk? Does the local stream meet state water quality standards?

Studies conducted by the USEPA to determine the correlation between different bacterial indicators and the occurrence of digestive system illness at swimming beaches suggest that the best indicators of health risk from recreational water contact in freshwater are *E. coli* and enterococci. For salt water, enterococci are the best. Interestingly, fecal coliforms as a group were determined to be a poor indicator of the risk of digestive system illness. However, many states continue to use fecal coliforms as their primary health risk indicator.

If your state is still using the total of fecal coliforms measurement as the indicator bacteria, and you want to know whether the water meets state water quality standards, you should monitor fecal coliforms. However, if you want to know the health risk from recreational water contact, the results of the USEPA studies suggest that you should consider switching to the *E. coli* or enterococci method for testing freshwater. In any case, consulting with the water quality division of your state's environmental agency is best, especially if you expect to use your data.

10.7.12.3 Sampling and Equipment Considerations

Bacteria can be difficult to sample and analyze, for many reasons. Natural bacteria levels in streams can vary significantly; bacteria conditions are strongly correlated with rainfall, making the comparison of wet and dry weather bacteria data a problem; many analytical methods have a low level of precision, yet can be quite complex to accomplish; and absolutely sterile conditions are essential to maintain while collecting and handling samples.

The primary equipment decision to make when sampling for bacteria is what type and size of sample container you will use. Once you have made that decision, the same straightforward collection procedure is used, regardless of the type of bacteria being monitored.

When monitoring bacteria, it is critical that all containers and surfaces with which the sample will come into contact be sterile. Containers made of either some form of plastic or Pyrex glass are acceptable to the USEPA. However, if the containers are to be reused, they must be sturdy enough to survive sterilization using heat and pressure. The containers can be sterilized by using an autoclave, a machine that sterilizes with pressurized steam. If using an autoclave, the container material must be able to withstand high temperatures and pressure. Plastic containers—either high-density polyethylene or polypropylene—might be preferable to glass from a practical standpoint because they will better withstand breakage. In any case, be sure to check the manufacturer's specifications to see whether the container can withstand 15 minutes in an autoclave at a temperature of 121°C without melting. (Extreme caution is advised when working with an autoclave.) Disposable, sterile, plastic Whirl-pak® bags are used by a number of programs. The size of the container depends on the sample amount needed for the bacteria analysis method you choose and the amount needed for other analyses.

The two basic methods for analyzing water samples for bacteria in common use are the membrane filtration and multiple tube fermentation methods (see Sections 6.3.6.3 and 6.3.6.4).

Given the complexity of the analysis procedures and the equipment required, field analysis of bacteria is not recommended. Bacteria can either be analyzed by the volunteer at a well-equipped lab or sent to a state-certified lab for analysis. If you send a bacteria sample to a private lab, make sure that the lab is certified by the state for bacteria analysis. Consider state water quality labs, university and college labs, private labs, wastewater treatment plant labs, and hospitals. You might need to pay these labs for analysis.

This text does not address laboratory methods, because several bacteria types are commonly monitored and the methods are different for each type. For more information on laboratory methods, refer to *Standard Methods*.

Note: If you decide to analyze your samples in your own lab, be sure to carry out a quality assurance/quality control program.

10.7.13 APPARENT COLOR

As we pointed out earlier in Sections 7.3 and 10.6.3, some aspects of water quality can be judged by color. Noticeable color is an objectionable characteristic that makes the water psychologically unacceptable to the consumer (De Zuane, 1997).

Pure water is colorless, but water in nature is often colored by foreign substances. Water with color is partly due to dissolved solids that remain after removal of suspended matter; this color is known as *true color*. *Apparent color* (the topic of this section) results from dissolved substances and suspended matter, and provides useful information about the water's source and content. Simply stated, when turbidity is present, so is apparent color. Natural metallic ions, plankton, algae, industrial pollution, and plant pigments from humus and peat may all produce color in water. Pure water absorbs different wavelengths (colors) of light at different rates. Blue and blue-green light are the wavelengths which are best transmitted through water, so a white surface under "colorless" water looks blue (e.g., Caribbean and some South Pacific Island waters above white sand).

Over the years, several attempts to standardize the method of describing the "apparent" color of water using comparisons to color standards have been made. *Standard Methods* recognizes the Visual Comparison Method as a reliable method of analyzing water from the distribution system.

One of the visual comparison methods is the Forel-Ule Color Scale, consisting of a dozen shades ranging from deep blue to khaki green, typical of offshore and coastal bay waters. By using established color standards, people in different areas can compare test results.

Another visual comparison method is the Borger Color System, which provides an inexpensive, portable color reference for shades typically found in natural waters; it can also be used for its original purpose—describing the colors of insects and larvae found in streams or lakes. The Borger Color System also allows the recording of the color of algae and bacteria on stream beds.

Note: Do not leave color standard charts and comparators in direct sunlight.

Measured levels of color in water can serve as indicators for a number of conditions. For example, transparent water with a low accumulation of dissolved minerals and particulate matter usually appears blue, and indicates low productivity. Yellow to brown color normally indicates that the water contains dissolved organic materials, humic substances from soil, peat, or decaying plant material. Deeper yellow to reddish colors in water indicate some algae and dinoflagellates in the water. Water rich in phytoplankton and other algae appears green. A variety of yellows, reds, browns, and grays are indicative of soil runoff.

To ensure reliable and accurate descriptions of apparent color, use a system of color comparison that is reproducible and comparable to the systems used by other groups.

10.7.14 ODOR

Odor in water (see Section 7.2) is caused by chemicals that may come from municipal and industrial waste discharges, or natural sources such as decomposing vegetable matter or microbial activity. Odor affects the acceptability of drinking water, the aesthetics of recreation water, and the taste of aquatic foodstuffs.

A wide variety of smells can be accurately detected by the human nose, which is the best odor-detection and testing device presently available. To measure odor, collect a sample in a large-mouthed jar. After waving off the air above the water sample with your hand, smell the sample. Use the list of odors provided in *Standard Methods* (a system of qualitative description that helps monitors describe and record detected odors; see Table 10.6) to describe the smells. Record all observations.

TABLE 10.6. **Descriptive System.**

Nature of Odor	Description	Such as Odors of
Aromatic	spicy	camphor, cloves, lavender, lemon
Balsamic	flowery	geranium, violet, vanilla
Chemical	industrial wastes or treatments	
	chlorinous	chlorine
	hydrocarbon	oil refinery wastes
	medicinal	phenol and iodine
	sulfur	hydrogen sulfide (rotten egg)
Disagreeable	fishy	dead algae
	pigpen	algae
	septic	stale sewage
Earthy	damp earth	
	peaty	peat
Grassy	crushed grass	
Musty	decomposing straw	
	moldy	damp cellar
Vegetable	root vegetables	

Source: Adapted from *Standard Methods*, 20th ed.

10.8 SUMMARY

All of the elements that comprise the standard practices associated with proper water monitoring provide drinking water practitioners with the technical and scientific data needed to determine the level and types of treatment needed to successfully condition the water obtained from surface and groundwater sources. We cover drinking water treatment in Chapter 11.

10.9 REFERENCES

AWWA. *Water Treatment*, 2nd ed. Denver: American Water Works Association, 1995.

De Zuane, J., *Handbook of Drinking Water Quality*. New York: John Wiley and Sons, Inc., 1997.

Einstein, A., Speech at the California Institute of Technology, California, February 16, 1931.

The Monitor's Handbook. Chestertown, Maryland: LaMotte Co., 1992.

River Watch Network. *Total Alkalinity and pH Field and Laboratory Procedures.* University of Massachusetts Water Resources Research Center: Acid Rain Monitoring Project, July 1, 1992.

Standard Methods for the Examination of Water and Wastewater. Washington, D.C: APHA (American Public Health Association), current edition.

USEPA. *Test Methods for* Escherichia coli *and Enterococci in Water by the Membrane Filter Procedure* (Method #1103.1). EPA 600/4-85-076. Cincinnati, OH: U.S. Environmental Protection Agency, Office of Research and Development, 1985.

USEPA. *Bacteriological Ambient Water Quality Criteria for Marine and Fresh Recreational Waters.* EPA 440/5-84-002. Cincinnati, OH: U.S. Environmental Protection Agency, Office of Research and Development, 1986.

USEPA. *Methods for Chemical Analysis of Water and Wastes,* 2nd ed. Washington, D.C.: U.S. Environmental Protection Agency, 1991.

USEPA. *Volunteer Stream Monitoring: A Methods Manual,* 2.841B97003. USEPA, 1997.

CHAPTER 11

Water Treatment

On Dec. 3, 1998, U.S. President Bill Clinton visited a water treatment plant in Newport, R.I., to announce new measures that will strengthen drinking water protection for at least 140 million Americans.

The new standards are a critical milestone on the path toward a new model for solving complex, high-stakes environmental problems. Where a comprehensive solution could be extremely costly and require essential information that is currently unavailable, the best approach may be to make decisions in stages. The initial decision may be a commitment to a process that will result in a comprehensive answer, starting with information-gathering, analysis and research, and accompanied by a broad set of consensus, "no regrets" actions. Under such a model, however, taking all the initial actions, as we have now done, is vital to sustaining the good-faith commitment to complete the process. This is such a promising model that it is being applied to address the highly controversial problems . . . and could have still broader applications. This one useful step for safe drinking water could become a giant leap for protection of public health and the environment. (J. C. Fox, 1999, p. 4)

11.1 INTRODUCTION

WE live on a planet with a surface that is three fourths covered with water, so we recognize the irony inherent in the fact that many areas of the world face critical shortages of drinking water. Most of Earth's water is seawater—far too saline for human consumption. Of the little "fresh" water that remains, most is trapped in polar ice caps where harnessing it for use by the world's population is difficult.

Much of the accessible natural supply of potable water faces stresses from a growing world population, which increases the basic demand for this natural resource while reducing the supply further through contamination.

Major population centers in developing nations (those without established waste treatment or water treatment infrastructures) often suffer from epidemics of waterborne disease. In these areas, raw sewage often directly contaminates the rivers and streams used for drinking, washing, and cooking. In other cases, unchecked industrialization leads to water contamination through improperly disposed of chemical and nuclear wastes.

Thus the drinking water purveyor must ensure that the drinking water supplied is safe for human consumption. In fact, the primary reason for the development and installation of a public water system is the protection of public health. Basically, a properly operated water system serves as a line of defense between disease and the public.

Properly operated water treatment and supply systems are defined as those that:

- remove or inactivate pathogenic microorganisms including bacteria, viruses, and protozoans
- reduce and/or remove chemicals that can be detrimental to health
- provide quality water thereby discouraging the customer from seeking better tasting or better looking water that may be contaminated

The last point above is critical but one often overlooked in the operation and management of public water systems. When the water produced by a system is objectionable because of odor, taste, or appearance, customers will often seek another source for their drinking water. Ironically, these sources, while appearing, tasting, and smelling fine (better than "city water"), could contain microorganisms and/or chemicals that are harmful.

In this chapter, we discuss the drinking water practitioner's most important function: ensuring that water delivered to the public is properly treated and arrives as the clean, wholesome, safe product that it must be.

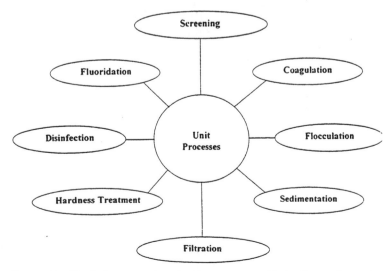

Figure 11.1 Typical water treatment unit processes used in treating surface waters.

A typical water treatment plant (treating stream or river water—turbid surface water with organics) processes raw water using various unit processes including: (1) screening, (2) coagulation, (3) flocculation, (4) sedimentation or settling, (5) filtration, (6) hardness treatment, (7) disinfection, and (8) fluoridation (see Figure 11.1). This chapter provides a brief overview of each of these unit processes that constitute a typical drinking water treatment system for surface water supplies.

11.2 SCREENING

In Section 5.2.1, we discussed water intakes and some of the included screening devices. *Screening* (the first important step in treating water containing large solids) is defined as the process whereby relatively large and suspended debris is removed from the water before it enters the plant. River water (the source of water used in our discussion) frequently contains suspended and floating debris varying in size from small rags to logs. Removing these solids is important, not only because these items have no place in potable water, but also because this river trash may cause damage to downstream equipment (clogging and damaging pumps, etc.), increase chemical requirements, impede hydraulic flow in open channels or pipes, or hinder the treatment process (Pankratz, 1995). The most important criteria used in the selection of a particular screening system for water treatment technology are the screen opening size and flow rate. Other important criteria include: costs related to operation and equipment; plant hydraulics; debris handling requirements; and operator qualifications and availability.

Large surface water treatment plants may employ a variety of screening devices including trash screens (or trash rakes), traveling water screens, drum screens, bar screens, or passive screens. We briefly discuss each of these screening devices in the following sections.

11.2.1 TRASH SCREENS (RAKES)

Trash screens or *trash rakes* are used to remove rough or large debris retained on a trash rack. They protect pumping equipment and may be used as a preliminary screening device to protect finer screens (drum or traveling water screens, for example).

A trash screen consists of one or more stationary trash rakes and a screen raking device. Trash rack bar spacings range from 1.5 to 4 inches and are mostly constructed of steel bars. Those constructed of high density polyethylene polymers are beginning to replace many of the older steel bar models; these synthetic screens are lighter and less prone to microbial growth, corrosion, and ice.

Raking mechanisms are available for use in a variety of intake configurations, including installation on vertical building and dam walls. Rakes are typically mounted either on fixed structures designed to clean a single trash rack, wheel-mounted to traverse the entire width of an intake structure and clean individual sections of a wide trash rack, or suspended from an overhead gantry.

11.2.2 TRAVELING WATER SCREENS

Traveling water screens (sometimes called bandscreens) are placed in a channel of flowing water to remove floating or suspended debris. These automatically cleaned screening devices protect pumping or other downstream equipment from debris in surface water intakes.

Consisting of a continuous series of wire mesh panels bolted to basket frames or trays, and attached to two matched strands of roller chain, the traveling water screen operates in a vertical path over a sprocket assembly through the flow. As raw water passes through the revolving baskets, debris is collected and retained on the upstream face of the wire mesh panels. The debris-laden baskets are lifted out of the flow and above the operating flow, where a high-pressure water spray directed outward removes the impinged debris. This process can be continuous or intermittent. For intermittent operation, the screen activates when a specified headloss or time elapsed has occurred.

When located on a river, traveling water screens may be subject to large fluctuations in flow conditions, debris loading, water depths, and salinity. Depending upon application, the traveling water screen's size is determined by considering such factors as maximum-average flow, maximum-minimum-average water levels, wire mesh size, velocity through mesh, basket or channel width, number of screens, type of service, and/or starting/operating headloss requirements.

11.2.3 DRUM SCREENS

A *drum screen* (or cylinder screen) has very few moving parts and is mounted on a horizontal axis with a series of wire mesh panels attached or mounted on the periphery of its cylinder. The cylinder slowly rotates on its axis. Because of its simplicity of construction, the maintenance and operating costs of a drum screen installation are usually less than those of a traveling water screen.

11.2.4 BAR SCREENS

Though primarily used in wastewater treatment applications, bar screens are also employed in some water treatment facilities. A *bar screen* consists of straight steel bars welded at both ends to two horizontal steel members, and is automatically cleaned by one or more power operated rakes. As a rake is operated up the face of the bar rack, it removes accumulated debris (usually large solid objects and rags) and elevates in and out of the flow. At the top of the rake's operating cycle, the debris is swept from the rake into a debris receptacle by a wiper mechanism. When installed in a waterway, the bar screen assembly normally is placed at an angle of 60–80° from the horizontal.

11.2.5 PASSIVE SCREENS

Passive intake screens (stationary screening cylinders) have no moving parts and require no debris handling or debris removal equipment. Passive intake screens are placed in a surface water body in such a manner so as to take advantage of natural ambient currents and controlled through-screen velocities to minimize debris buildup.

Usually mounted on a horizontal axis and oriented parallel to the natural current flow within the water body, current flow action works to keep the screen clean. Maximum intake velocity of about 0.5 fps is typical and works to minimize debris impingement on the screen surface.

11.3 COAGULATION

Coagulation, the second step in water purification, is a unit process that has been used for several years in the treatment of raw water. Basically, coagulation works to settle very fine material of suspended solids.

Note: Chemicals employed for coagulation are expected to be safe for drinking water, when used according to standards.[18]

11.3.1 THE COAGULATION PROCESS

Typically, after screening, raw water is pumped into large *settling basins* (also known as clarifiers or sedimentation tanks). Within the confines of the settling basin, the screened raw water is allowed to sit for some predetermined time. Although screened, the raw water still contains impurities that may be either dissolved or suspended. The settling basin provides the most convenient way in which to remove the suspended matter: it lets the force of gravity do the work. Within the basin, when flow and turbulence are minimal (quiescent conditions), particles more dense than water settle to the bottom of the tank. This process is called *sedimentation*, and the layer of

---

[18]*AWWA Standards* (Coagulation—Nos. 42402 to 42407), AWWA, Denver, Colorado, latest edition.

accumulated solids at the bottom of the tank is called *sludge* (or *biosolids* in some wastewater treatment unit processes).

The size and density of the suspended particles has a direct bearing on the speed at which they will settle toward the bottom of the basin. The larger and heavier particles will, of course, settle faster than smaller or lighter particles. The forces opposing the downward force of gravity include buoyancy and drag (friction). Particle-settling rate is also affected by the temperature and viscosity of the water.

Note: The nature of the sedimentation process also varies with the concentration of suspended solids and their tendency to interact with one another.

In the sedimentation process just described, not all suspended solids or particles can be completely removed from water, even when given very long detention times. For example, very small particles called *colloids* (and other particles such as bacteria, particles of color, and turbidity, which may be colloidal), will not settle out of suspension by gravity without some help. This is where *coagulants* come into play. If we rapidly mix chemical coagulants in the water, then slowly stir the mixture before allowing sedimentation to occur, the colloidal particles will settle.

Colloids or finer particles must be chemically coagulated to produce larger floc removable in subsequent settling and filtration.

The coagulation process (along with flocculation) works to neutralize or reduce the natural repelling electrical force of particles in water, keeping them apart and in suspension. Particles in water usually carry a negative electrical charge. Since all these particles carry this same negative electrical charge, they repel each other—in the same way that poles of a magnet do.

The object of coagulation (and subsequently flocculation) is to turn the small particles into larger flocs, either as precipitates or suspended particles. These flocs are then conditioned for ready removal in subsequent processes. In this text we define *coagulation* as a method to alter the colloids so that they will be able to approach and adhere to each other to form larger floc particles.

11.3.2 COAGULANTS

Coagulants and coagulant aids are used in the coagulation process. Generally, the types of coagulants and aids available are defined by the plant process scheme. To determine optimum chemical dosages for treatment, jar tests are normally used.

11.3.2.1 Jar Test

Jar tests are widely used to simulate a full-scale coagulation (and flocculation) process to determine optimum chemical dosages (the cost-effective dose of a coagulant for the time and intensity of agitation selected). Used for many years by the water treatment industry, the test conditions are intended to reflect the normal operation of a chemical treatment facility. In this way, the type and quantity of sludge and the physical properties of the floc can be evaluated.

The test can be used to:

- select the most effective chemical
- select the optimum dosage
- determine the value of a flocculant aid and the proper dose

The testing procedure requires a series of samples to be placed in testing jars and mixed at 100 rpm. Varying amounts of the process chemical or specified amounts of several flocculants are added (one volume/sample container). The mix is continued for one minute. The mixing slows to 30 rpms to provide gentle agitation, then the floc is allowed to settle. The flocculation period and settling process is observed carefully to determine the floc strength, settleability, and clarity of the supernatant liquor (the water that remains above the settled floc). The supernatant can then be tested to determine the efficiency of the chemical addition for removal of TSS, BOD_5, and phosphorus.

The equipment required for the jar test includes a six-position variable speed paddle mixer, six two-quart wide mouth jars, an interval timer, and assorted glassware, pipettes, graduates, and so forth.

11.3.3 TYPES OF COAGULANTS

Several different chemicals can be used for coagulation. Commonly used metal coagulants are those based on aluminum (aluminum sulfate) and those based on iron (ferric sulfate). The most common coagulant is aluminum sulfate [also known as alum, $Al_2(SO_4)_3$]. We list other common coagulation chemicals in Table 11.1.

11.3.4 TYPES OF COAGULANT AIDS

Difficulties with coagulation often occur because of slow-settling precipitates or fragile flocs that are easily fragmented under hydraulic forces in basins and filters (Hammer and Hammer, 1996). AWWA (1995) describes a coagulant aid as a chemical added during coagulation to improve coagulation; to build stronger, more settleable floc; to overcome the effect of temperature drops that slow coagulation; to reduce the amount of coagulant needed; and to reduce the amount of sludge produced. Coagulant aids benefit flocculation by improving the settling qualities and toughness of flocs.

TABLE 11.1. Common Coagulant Chemicals.

Common Name	Comments
Aluminum sulfate	Most common coagulant in the United States; often used with cationic polymers
Ferric chloride	May be more effective than alum in some applications
Ferric sulfate	Often used with lime softening
Ferrous sulfate	Less pH dependent than alum
Aluminum polymers	Synthetic polyelectrolytes; large molecules
Sodium aluminate	Used with alum to improve coagulation
Sodium silicate	Ingredient of activated silica coagulant aids

Source: Adapted from Water Treatment Plant Design, Larsen, p. 77, 1990.

Polymers are the most widely used materials. Synthetic polymers are water-soluble, high-molecular weight organic compounds with multiple electrical charges along a molecular chain of carbon atoms. In drinking water treatment, polymers are extensively used as coagulant aids to build large floc prior to sedimentation and filtration. Other coagulant aids are activated silica, adsorbent-weighting agents, and oxidants.

11.3.5 PROCESS OPERATION

The common coagulation unit process operation involves the addition of coagulant chemicals by rapid mixing; detention time in the rapid mix tank is typically on the order of minutes (Masters, 1991). During this mixing process, polymer (or some other coagulant aid) is added and blended into the destabilized water prior to and during flocculation.

The removal of impurities by coagulation depends on their nature and concentration; use of both coagulants and coagulant aids; and other characteristics of the water, including pH, temperature, and ionic strength. Because of the complex nature of coagulation reactions, chemical treatment is based on empirical data derived from jar testing or other laboratory tests and field studies (Viessman and Hammer, 1998).

11.4 FLOCCULATION

The destabilized particles and chemical precipitates resulting from coagulation are designed to enhance their settling qualities and thus their removal from water. However, even after coagulation has taken place, these particles and chemical precipitates may still settle very slowly (too slowly). To speed up the settling process, the next step in the water treatment process, flocculation, is employed.

Note: Flocculation is the "clumping" of coagulated particles as the result of coagulation. While the two terms are often used interchangeably, they are actually distinct concepts.

Flocculation is the most important factor affecting particle-removal efficiency. In operation, flocculation is a slow mixing process in which these particles are brought into contact so that they will collide, stick together, and grow (agglomerate) to a size that will readily settle. Enough mixing must be provided (gentle agitation for approximately one-half hour) to bring the floc particles into contact with each other, and to keep the floc from settling in the flocculation basin. (The heavier the floc and the higher the suspended solids concentration, the more mixing is required to keep the floc in suspension.) The most common types of "mixer" or flocculator are paddle types, which use redwood slats mounted horizontally on motor-driven shafts. Rotating slowly at about one revolution per minute, the paddles provide gentle agitation that promotes floc growth. The rate of agglomeration or flocculation depends on the number of particles present, the relative volume which they occupy, and the velocity gradient in the basin.

Note: When we state that the rate of agglomeration or flocculation depends on *velocity gradient*, we refer to the fact that too much mixing shears the floc particles, tearing them apart again, so that the floc becomes small and finely dispersed (a situation we are obviously trying to avoid). Thus, the velocity gradient must be controlled within a relatively narrow range.

The theory of flocculation is complex and beyond the needs of this text, but on an elemental level we can say that flocculation is generally effected by slowly rotating large-diameter mixers. Recent practice incorporates dispersion of the coagulant (flash mixing), flocculation, and sedimentation in a single unit called a *contact clarifier*.

Note: Flocculation is a principal mechanism in removing turbidity from water.

11.5 SEDIMENTATION

In a conventional water treatment plant, coagulation/flocculation precedes the sedimentation process for better results and improved utilization of the settling basins, and then is followed by the filtration process. In the past (but not necessarily confined to the past), filtration was preceded only by coagulation; here filtration was provided only after a few minutes of contact, with consequent additional stress to the filters. Lack of sedimentation means less reliability on operation of filters, when water quality suddenly changes characteristics (De Zuane, 1997).

Sedimentation (also known as *clarification*) is the gravity-induced removal of particulate matter, chemical floc, precipitates from suspension, and other settleable solids. Simply stated, sedimentation separates the liquid from the solids. The process takes place in a rectangular, square, or round tank called a settling or sedimentation tank or basin. Flow patterns within such basins may be rectilinear flow in rectangular basins; radial flow in center-feed settling tanks; radial flow in peripheral-feed settling tanks; spiral flow in peripheral-feed settling tanks; or radial flow in square settling tanks.

Sedimentation, in the conventional water treatment process, is typically the step between flocculation and filtration. Design criteria are based on empirical data from the performance of full-scale sedimentation tanks. The common criteria for sizing settling basins are detention time (typically from 1 to 10 hours); overflow rate; weir loading; and with rectangular tanks, horizontal velocity.

In water treatment, the majority of settling basins are essentially upflow clarifiers where the water rises vertically for discharge through effluent channels. More specifically, in the idealized sedimentation tank, water flows horizontally through the basin, then rises vertically, overflowing the weir of a discharge channel at the tank surface. Floc settles downward, opposite the upflow of water, and is removed from the bottom by a continuous mechanical sludge removal apparatus. The particles with a settling velocity greater than the overflow rate are removed (settled) while lighter flocs are carried out in the effluent. The effluent is then filtered.

Note: Sedimentation basins, either circular or rectangular, are designed for slow uniform water movement with a minimum of short-circuiting.

11.6 FILTRATION

Even after chemical coagulation and sedimentation by gravity, not all of the suspended solids or impurities are removed from water. Nonsettleable floc particles (about 5% of the suspended solids) may still remain in the water. With only a small percentage left, we might ask if this weren't good enough. But it isn't. This remaining floc would cause problems (including noticeable turbidity); and particles shield microorganisms from the subsequent disinfection processes. The water treatment's goal is to produce potable water that is perceptually crystal clear and that satisfies the Safe Drinking Water Act requirement of 0.5 NTU (for turbidity). To accomplish this, water needs an additional treatment step that follows coagulation, flocculation, and sedimentation.

This next step is the physical process of *filtration*. Filtration (sometimes called a polishing process) involves the removal of suspended particles from water by passing it through a layer or *bed* of a porous granular material—sand, for example. As water flows through the filter bed, the suspended particles become trapped within the pore spaces of the filter material (or *filter media*). In purifying a surface water source (as we are discussing in this case), filtration is a very important process, even though filtration is only one step in the overall treatment process.

Note: Filtration is the process that occurs naturally as surface waters migrate (percolate) through the porous layers of soil to recharge groundwater. This natural filtration removes most suspended matter and microorganisms and is the reason many wells produce water that doesn't require any further treatment.

The Surface Water Treatment Rule (SWTR) specifies certain filtration technologies. The most common treatment filter systems include rapid gravity filters, either built on-site or packaged plants, and pressure filters. Another type is direct filtration. Two other types are slow sand filters and the diatomaceous earth (DE) filters. The SWTR allows the use of "alternate" filtration technologies, such as cartridge filters.

Filtration treatment unit processes most commonly used in water purification systems include slow or rapid sand filtration, diatomaceous earth filtration, and package filtration systems. Slow and rapid filter systems refer to the rate of flow per unit of surface area. Filters are also classified by the type of granular material used in them. Sand, anthracite coal, coal-sand, multilayered, mixed bed, or diatomaceous earth are examples of different filtering media. Filtration systems may also be classified by the direction the water flows through the medium. Some examples are downflow, upflow, fine-to-coarse, or coarse-to-fine. Lastly, filters are commonly distinguished by whether they are gravity or pressure filters. Gravity filters rely only on the force of gravity to move the water down through the grains and typically use upflow for washing (backwashing) the filter media to remove the collected foreign material. Gravity filters are free surface filters that are much more commonly used for municipal applications. Pressure filters are completely enclosed in a shell, so that most of the water pressure in the lines leading to the filter is not lost and is used to push the water through the filter.

11.6.1 RAPID FILTER SYSTEMS

Slow fine sand filtration has been used in the U.S. since 1872; although still used in many older plants, it is not commonly used today in most modern water treatment plants, because of various problems associated with it. One of the problems is related to the tiny size of the pore spaces in the fine sand, which slows down the water, slowing its progress through the filter bed. These filter types also have problems with suspended particles clogging the surface, requiring the filter to be manually scraped clean. And these units take up a considerable amount of land area, because slow filtration rates mean higher filter surface area to produce the needed filtered water qualities.

In modern water treatment plants, the *rapid filter* has largely replaced the slow sand filter. The rapid filter consists of a layer of carefully sieved silica sand ranging from 0.6 to 0.75 m in depth on top of a bed of graded gravels. The pore openings between the grains of sand are often greater than the size of the floc particles that are to be removed, so much of the filtration is accomplished by means other than simple straining.

Note: The ideal filter media possesses the following characteristics: coarse enough for large pore openings to retain large quantities of floc, yet sufficiently fine to prevent passage of suspended solids; adequate depth to allow relatively long filter runs; and graded to permit effective cleaning during backwash.

Adsorption, continued flocculation, and sedimentation in the pore spaces are also important removal mechanisms. When the filter becomes clogged with particles (which occurs approximately once a day, depending on the turbidity of the water), the filter is shut down for a short period of time and cleaned by forcing water backwards through the sand for 10–15 minutes. After cleaning, the sand settles back in place and operation resumes.

11.6.2 OTHER COMMON FILTER TYPES

Rapid-flow filters are the most common type used for treating water supplies, primarily because they are the most reliable. But other types of filters are sometimes used to clarify water, including pressure filters and diatomaceous earth filters.

A *pressure filter* is similar to a rapid filter in that the water flows through a granular filter bed. However, instead of being open to the atmosphere and using the force of gravity, the pressure filter is enclosed in a cylindrical steel tank and the water is pumped through the bed under pressure. Pressure filters are not as reliable as rapid filters, because pressure may force solids through the bed in the effluent. Because of this problem, they are seldom employed in municipal water treatment works, but instead are used for filtering water for industrial use or in swimming pools.

Diatomaceous earth filters contain a thin layer of a natural powderlike material formed from the shells of diatoms, and are also used primarily for industrial or swimming pool applications because they are not as reliable as rapid sand filters.

11.7 HARDNESS TREATMENT

Two commonly used methods to reduce hardness are the lime-soda process and ion exchange. The lime-soda process is applicable for large facilities, while ion exchange is normally employed in smaller waterworks. The lime-soda process will not remove all of the hardness and is usually operated to produce a residual hardness of about 100 mg/L as $CaCO_3$.

Greater reductions are not economical and may have adverse health consequences as well (McGhee, 1991). In this text, our discussion focuses on ion
exchange.

Ion exchange is accomplished by charging a resin with sodium ions and
allowing the resin to exchange the sodium ions for calcium and/or magnesium ions.

Common resins include zeolites—natural and man-made minerals that
will collect from a solution certain ions (sodium or $KMnO_4$), and either
exchange these ions (the case in water softening), or use the ions to oxidize
a substance (in the case of iron or manganese removal).

The negative side of using ion exchange is that even though the process
softens water by removing all (or nearly all) of the hardness and adds sodium
ions to the water, the water may be more corrosive than before. The addition
of sodium ions to the water may also increase the health risk of those with
high blood pressure.

11.8 DISINFECTION

At the turn of the century, 35,000 people per 1,000,000 people did not
reach 20 years of age. Today, according to *USA TODAY* (1998), while 101
deaths occur every minute, at the same time, 261 births occur. Curbing
waterborne disease through disinfection has significantly contributed to
birthrates outpacing deathrates worldwide.

The Safe Drinking Water Act requires that public water supplies be
disinfected, and that the USEPA set standards and establish processes for
treatment and distribution of disinfected water, to ensure that no significant
risks to human health occur. The USEPA Science Advisory Board ranks
pollutants in drinking water as one of the highest health risks meriting
USEPA's attention, because of large-scale population exposure to contaminants, including lead, disinfectant by-products (DBPs), and disease-causing
organisms. Disinfectants are used by virtually all surface water systems in
the U.S. and by an unknown percentage of systems that rely on groundwater.
For nearly a century, chlorine has been the most widely used and most cost
effective disinfectant. However, disinfection treatments can produce a wide
variety of by-products, many of which have been shown to cause cancer, or
that can produce other toxic effects. Recently, concern has been raised over
water quality deterioration, which can increase dramatically during distribution, unless systems are properly designed and operated. While disinfection is an integral part of water treatment, filtration prior to disinfection is
necessary to reduce pathogen levels, making disinfection more reliable by
removing turbidity and other interfering constituents.

To solve the disinfection by-products problem, we need innovation to upgrade existing techniques, as well as to develop new approaches to address these problems. Areas of interest include:

- alternatives to chlorine disinfection for removing pathogenic microorganisms, including innovative applications of ultraviolet radiation and processes that improve overall effectiveness while using reduced amounts of disinfectant
- development of innovative unit processes, particularly for small systems, for removal of organic and inorganic contaminants (such as arsenic), particulates, and pathogens [e.g., cyst-like organisms and emerging pathogens like caliciviruses, microsplorida (septata and enterocytozoan), hepatitis A virus (HAV), *Mycobacterium avium* intracellulare (MAC), *Helicobacter pylori, Lengioneela pneumophilar,* adenovirus 40/41/1–39, and *Toxoplasma gondii*]
- development of efficient, cost-effective treatment processes for removing disinfection by-product precursors (e.g., trihalomethanes, haloacetic acids; and for ozonation: bromate, aldehydes; for chlorination: chloropicrin, haloacetonitriles; for chloramination: organic chloramines, cyanogen chloride)
- improved methods for controlling pathogens through coagulation/settling, filtration, or other cost-effective means
- drinking water contamination control between the treatment plant and the user, especially considering potential chemical leaching from distribution system materials and surfaces (e.g., lead, copper, iron and other pipe materials, protective coatings) as a result of instability, interaction with microorganisms, disinfection agents, and water treatment chemicals

To this point, the unit processes described—screening, coagulation, flocculation, sedimentation, and filtration—together comprise a type of treatment called *clarification*. Along with removing turbidity and suspended solids, clarification also removes many microorganisms from the water. However, clarification by itself is not sufficient to ensure the complete removal of pathogenic bacteria and viruses.

Earlier we stated that one of the primary goals of water treatment is to treat raw water to the point where we can deliver to the consumer a water product that is perceptually crystal clear. Obviously the consumer does not want to drink a glass full of mud, a glass full of slime, a glass full of metal-colored, foul-smelling water, or even a glass of water that looks like it was dipped from the creek.

However, once the water is treated to the point of crystal clarity, the treatment process must still be taken a step further—to the point where the water is completely free of disease-causing microorganisms. To accomplish this, the final treatment process in water treatment plants occurs—*disinfection*, which destroys or inactivates pathogens.

11.8.1 KEY DISINFECTION TERMS[19]

Before we move on to a discussion of the major disinfection methods used in treating water for human consumption, we must first define a few pertinent terms related to disinfection in general. To begin with, we need to establish the distinction between primary and secondary disinfection.

(1) *Primary Disinfection:* the initial killing of *Giardia* cysts, bacteria, and viruses.

(2) *Secondary Disinfection:* the maintenance of a disinfectant residual, which prevents regrowth of microorganisms in the water distribution system between treatment and consumer.

General disinfection terms include the following:

- *Disinfect*—to inactivate virtually all recognized pathogenic microorganisms, but not necessarily all microbial life (compare to pasteurize or sterilize).
- *Disinfectant*—(1) any oxidant, including but not limited to chlorine, chlorine dioxide, chloramine, and ozone, added to water in any part of the treatment or distribution process that is intended to kill or inactivate pathogenic microorganisms; (2) a chemical or physical process that kills pathogenic organisms in water. Chlorine is often used to disinfect sewage treatment effluent, water supplies, wells, and swimming pools.
- *Disinfectant time*—the time water takes to move from one point of disinfectant application (or the previous point of residual disinfectant measurement) to a point before or at the point where the residual disinfectant is measured.
- *Disinfectant contact time (*T *in* CT *calculation)*—the time in minutes that water takes to move from the point of disinfectant application or the previous point of disinfection residual measurement to a point before or at the point where residual disinfectant concentration (C) is measured. Where only one C is measured, T is the time in minutes that water takes to move from the point of disinfectant application to a point before or where residual disinfectant concentration (C) is measured. Where more than one C is measured, T is:
  (1) For the first measurement of C—the time in minutes that water takes

[19]Adapted from F. R. Spellman, *Disinfection Alternatives.* Lancaster, PA: Technomic Publishing Company, Inc., 1999.

to move from the first or only point of disinfectant application to a point before or at the point where the first C is measured.

(2) For subsequent measurements of C—the time in minutes that water takes to move from the previous C measurement point to the C measurement point for which the particular T is being calculated.

- *Disinfection*—a process which inactivates pathogenic organisms in water by chemical oxidants or equivalent agents.
- *Disinfection by-product*—a compound formed by the reaction of a disinfectant such as chlorine with organic material in the water supply.
- *Waterborne disease*—caused by pathogenic organisms in water.

Note: Disinfection should not be confused with sterilization.

Sterilization is the destruction of all microorganisms. Sterilizing potable water requires the application of a much higher dose of chemical disinfectants, which would greatly increase operating costs and would create taste problems for the consumer. Excessive application of disinfectants also generates excessive levels of unwanted disinfection by-products. Because of these problems, present treatment practices are used for turbidity removal and subsequent disinfection to the extent necessary to eliminate known disease-causing organisms sufficient to protect public health.

The presence of coliform bacteria in water is an indication that the water may be contaminated by pathogenic organisms. Absence of coliform bacteria is considered to be sufficient evidence that pathogens are absent—if the source is good, a chlorine residual level is maintained, and the supply has a good history.

11.8.2 DISINFECTION METHODS

Although chlorination is the best known and the most common disinfection method, other methods are available and can be used in various situations. The three general types of disinfection are:

- *Heat treatment*—employs boiling to disinfect water. Probably one of the first methods employed to disinfect water was to boil it. For small quantities of water, boiling is still a good emergency procedure to use.
- *Radiation treatment*—employs the use of ultraviolet (UV) radiation to disinfect water.
- *Chemical treatment*—employs the use of chemicals to disinfect water. Examples of chemical disinfectants include: oxidizing agents such as chlorine, ozone, bromine, iodine, potassium permanganate; metal ions such as silver, copper, mercury; and/or acids and alkalis.

Obviously, several different disinfectants are available for use in treating water, and we'll discuss some of these in detail in subsequent sections. For now, you should understand that even though several choices are available, whichever disinfectant is chosen must meet certain criteria. Specifically, to be effective for disinfecting water (and wastewater), the disinfectant chosen must possess certain desirable characteristics.

Desirable characteristics of a disinfectant include:

(1) Must act in a reasonable time
(2) Must act as temperature or pH changes
(3) Must be nontoxic
(4) Must not add unpleasant taste or odor
(5) Must be readily available
(6) Must be safe and easy to handle and apply
(7) Must be easy to determine the concentration of
(8) Must be able to provide residual protection
(9) Pathogenic organisms must be more sensitive to disinfectant than are nonpathogens
(10) Must be capable of being applied continually
(11) Amount applied must be sufficient to produce a safe water

In addition to having the characteristics listed above, the disinfectant chosen must be able to kill off or deactivate pathogenic microorganisms by one of several possible methods, including (1) damaging the cell wall; (2) altering the ability to pass food and waste through the cell membrane; (3) altering the cell protoplasm; (4) inhibiting the cell's conversion of food to energy; or (5) inhibiting reproduction.

11.8.2.1 Chlorination

For the past several decades, chlorine dispensed as either a solid (calcium hypochlorite), liquid (sodium hypochlorite), or gas (elemental chlorine, Cl_2) has been the disinfectant of choice, particularly in the United States. Chlorine (sometimes referred to as the workhorse of disinfection) has proven its worth both because of its effectiveness, and also because it is relatively inexpensive; it also provides a chlorine residual in the water distribution system, ensuring that the water remains disease-free.

Gaseous chlorine (Cl_2), 2.5 times as heavy as air, is a greenish-yellow toxic gas. One volume of liquid chlorine confined in a container under pressure yields about 450 volumes of gas. Large water treatment works usually use chlorine gas, supplied in liquid form, in high-strength, high-pressure steel cylinders. The liquid immediately vaporizes in the form of gas when released from these pressurized containers. Chlorine gas is lethal at concentrations as low as 0.1% air by volume. In nonlethal concentrations it irritates the eyes, nasal membranes, and respiratory tract.

Sodium hypochlorite is most commonly used in smaller systems, because it is simpler to use and has less extensive safety requirements than gaseous chlorine; in the form used, it is less toxic. Recently, many larger water facilities that have used chlorine for disinfection are beginning to substitute sodium hypochlorite for chlorine because of regulatory pressure.

Note: Two regulations [OSHA's Process Safety Management Standard, 29 CFR 1910.119 and USEPA's Risk Management Program, Clean Air Act Section 112(r)(7)] have come to be known in industry as the "chlorine killers," because of their effect on industrial processes. The USEPA is attempting to steer industry away from the use of chlorine. While they cannot absolutely outlaw this substance from use, they are following the path of simply regulating it to death. At present, that is what is occurring. To avoid compliance with strict (in some cases unworkable) regulations, substitution to some other chemical product that is not regulated (at least for the moment), such as sodium hypochlorite, is being effected by many water/wastewater facilities in the U.S.

Sodium hypochlorite provides 5% to 15% available chlorine. (Common laundry bleach is a five percent solution of sodium hypochlorite.) Usually diluted with water before application as a disinfectant, it is very corrosive and should be handled and stored with care and kept away from equipment that can be damaged by corrosion. Sodium hypochlorite solution is more costly per pound of available chlorine and does not provide the level of protection (the "killing" power against pathogens) of chlorine gas.

Calcium hypochlorite is a white solid in granular, powdered, or tablet form containing 65% available chlorine. In packaged form, calcium hypochlorite is stable, more stable than solutions of sodium hypochlorite, which deteriorate over time. However, calcium hypochlorite is *hygroscopic*, which means it readily absorbs moisture. It reacts slowly with moisture in the air to form chlorine gas. It is a corrosive material with a strong odor, and requires proper handling. Some practical difficulty is involved in dissolving calcium hypochlorite. It must be kept away from organic materials such as wood, cloth, and petroleum products. Reactions between it and organic materials can generate enough heat to cause a fire or explosion.

11.8.2.1.1 Chlorine Use

Whatever form of chlorine is used for disinfection (elemental chlorine, sodium hypochlorite, or calcium hypochlorite), it may be added to the incoming flow (prechlorination) to assist with the oxidation of inorganics or to arrest biological action that may produce undesirable gases in the bottom of clarifiers. More often, however, chlorine is added just prior to filtration to keep algae from growing at the medium surface and to prevent large populations of bacteria from developing within the filter medium. Safe and effective application of chlorine requires specialized equipment and considerable care and skill on the part of the plant operator. Various means of feeding chlorine have been developed, but probably one of the widest used and safest types of chlorine feed devices is called an *all-vacuum chlorinator*. Mounted directly on the chlorine cylinder, the gaseous chlorine is always under a partial vacuum in the line that carries it to the point of application. In a typical vacuum chlorine feed system, the vacuum is formed by water flowing through the ejector unit at high velocity.

Hypochlorites are usually applied to water in liquid form by means of positive displacement-type pumps, which deliver a specific amount of liquid on each stroke of a piston or flexible diaphragm. Chlorine, when added to water, reacts with various substances or impurities in the water (e.g., organic materials, sulfides, ferrous iron, and nitrites), creating a *chlorine demand*. Chlorine demand is a measure of the amount of chlorine that will combine with impurities and is therefore available to act as a disinfectant.

Chlorine combines with ammonia or other nitrogen compounds to form chlorine compounds that have some disinfectant properties. These compounds are called *combined available chlorine residual*. In the context used here, "available" means available to act as a disinfectant. The uncombined chlorine that remains in the water after combined residual is formed is called *free available chlorine residual*. Free chlorine is a much more effective disinfectant than combined chlorine.

11.8.2.1.2 Factors Affecting Successful Chlorination

The factors important to successful chlorination are

- concentration of free chlorine
- contact time
- temperature
- pH
- turbidity

The effectiveness of chlorination is directly related to the contact time with and concentration of free available chlorine. At lower chlorine concentrations, contact times must be increased. Maintaining a lower pH will also increase the effectiveness of disinfection. The higher the temperature, the faster the disinfection rate. Chlorine (or any other disinfectant for that matter) is effective only if it comes into *contact* with the organisms to be killed. Good contact between chlorine and microorganisms is prevented whenever high turbidity levels exist. For this and aesthetic reasons, turbidity should be reduced where necessary through the coagulation and sedimentation methods previously discussed.

11.8.3 CHLORINATION BY-PRODUCTS

A serious disadvantage of chlorination is the potential formation of by-products. Chlorine, for example, can mix with the organic compounds in water (such as decaying vegetation) to form trihalomethanes (THMs). One THM (chloroform) is a suspected carcinogen. Other common trihalomethanes are similar to chloroform and may cause cancer.

At the present time, on estimate, about 90% of U.S. water utilities use chlorine to disinfect water. Although chlorine has virtually eliminated the risks of waterborne diseases such as typhoid fever, cholera, and dysentery, recent studies have shown risks associated with by-products of chlorine—a reason why water utilities already have been looking at alternative methods for disinfecting water.

Several approaches for reducing harmful chlorination by-products have been used. For example, one approach is to remove more of the organics before any chlorination takes place. This can be accomplished (to a degree) by not chlorinating the incoming raw water before coagulation and filtration, thus reducing the formation of THMs. Aeration or adsorption on activated carbon will remove organic materials at higher concentrations or those not removed by other techniques. Another approach is to reevaluate the amount of chlorine used; the same degree of disinfection might be possible with lower chlorine dosages. Changing the point in treatment where chlorine is added is another approach commonly employed (i.e., instead of adding chlorine as chemical feed, during coagulation, sedimentation, or filtration, it can be added after filtration). Another current approach is using alternative disinfection methods.

Earlier we pointed out that because of OSHA's Process Safety Management Standard (PSM) and USEPA's Risk Management Program (RMP), many facilities currently using elemental chlorine have or are actively pursuing the use of alternative disinfection methods. We further re-emphasize that the problem of THMs is also helping spur interest in alternatives to chlorination as the preferred method of disinfection. We briefly discuss a few alternative disinfection methods in the following section.

11.9 ALTERNATIVE DISINFECTION METHODS

Currently several alternative disinfection methods are available for use in treating water. However, in this text (for water treatment) we focus on two alternatives, ozonation and ultraviolet (UV) radiation. These commonly used alternatives (especially in small water treatment systems) are also increasingly being substituted for existing chlorination systems at larger plants because of regulatory pressure.

Note: Before beginning a brief discussion of the ozonation and UV disinfection alternatives, we must point out that neither one of these two alternative disinfectants is an easy solution to problems created by chlorination. While true that each has the advantage of not creating THMs and not being covered by the requirements under PSM/RMP, it is also true that each has uncertainties and known disadvantages that have restricted their more widespread use. In addition, ozonation and UV irradiation cannot be used as disinfectants by themselves. Both require a secondary disinfectant (usually chlorine) to maintain a residual in the distribution system.

11.9.1 OZONATION

Ozone (O_3), a gas at ordinary temperature and pressures, is a very powerful disinfectant (it disinfects by breaking up molecules in water). It is even more effective against some viruses and cysts than chlorine. It has the added advantage of leaving no taste or odor and is unaffected by pH or the ammonia content of the water. When ozone reacts with reduced inorganic compounds and with organic material, an oxygen atom instead of a chloride atom is added to the organics, the end result being an environmentally acceptable compound. But since ozone is unstable and cannot be stored, it must be produced on-site. Ozonation usually costs more than chlorination as well.

11.9.2 UV

UV (ultraviolet) *light* is electromagnetic radiation just beyond the blue end of the light spectrum, outside the range of visible light. It has a much higher energy level than visible light, and in large doses it inactivates both bacteria and viruses. UV energy is absorbed by genetic material in the microorganisms, interfering with their ability to reproduce and survive, as long as the radiation contacts the microorganisms without interference from turbidity. The big advantage of UV disinfection over chlorine and ozone is that UV does not involve chemical use. Generally, UV light used for disinfecting water is generated by a series of submerged, low-pressure mercury lamps. Continuing advances in UV germicidal lamp technology are making UV disinfection a more reliable and economical option for disinfection in many plants.

11.10 NONCONVENTIONAL WATER TREATMENT TECHNOLOGIES

Stage 1 of the USEPA's Disinfectant/Disinfection By-Product Rule (D/DBP) and the new Interim Enhanced Surface Water Treatment Rule, designed to significantly lower trihalomethanes (by-products of chlorine disinfection) in water, have driven (along with the regulatory requirements of PSM/RMP) many water and wastewater utilities to find and use alternative disinfection methodologies. While ozonation and ultraviolet irradiation might be suitable disinfection alternatives, switching from chlorine to chlorine dioxide (a chemical which has been proven to form fewer THMs) might also be another viable disinfection alternative, and several others are possible as well. Which disinfection alternative is ultimately selected is driven not only by regulatory requirements, but also by site specific requirements.

The disinfection issues covered to this point are important because the overall ramifications of regulatory pressure and environmental impact cannot be overstated; but other issues besides disinfection must be considered when deciding which water treatment methodology to employ. For example, clarification by coagulation, flocculation, sedimentation, and filtration removes suspended impurities and turbidity from drinking water. Disinfection (the final step in the process) produces potable water, free of harmful pathogens. Simply put, the water treatment processes discussed in the previous sections of this part of the text are sufficient to render most natural surface water (such as the river source we have used for illustration) potable. In some instances, however, the water supply may contain materials that are not removed by conventional water treatment processes. Other treatment processes may be required, particularly to remove many of the dissolved organic and inorganic substances. For example, groundwater may contain excessive dissolved solids, and surface waters may contain organic compounds from domestic or industrial wastewaters or naturally occurring organics such as humic acid or products of algal blooms. Processes are available for removing these contaminants.

Note: These additional water treatment processes involve sophisticated equipment, require highly skilled operators, and are therefore quite expensive (Peavy et al., 1985).

Additional water unit treatment processes may be used in addition to clarification or applied separately, depending on the source and quality of the raw water. Let's take a closer look at groundwater. Does a typical groundwater source require treatment beyond conventional means? Groundwater does not normally require processing by the unit treatment steps listed above, other than disinfection, because groundwater is filtered naturally by the layers of soil from which it is withdrawn. Disinfection is only applied (in many cases) as a precautionary step required by law for public water systems. Groundwater is usually free of bacteria or other microorganisms. However, that groundwater comes into contact with soil and rock (and it all does) is a cause for concern. With such contact, groundwater may become contaminated by high levels of dissolved minerals that must be removed.

11.11 FLUORIDATION

Fluoride, when added to drinking water supplies in small concentrations (about 1.0 mg/L), can be beneficial. In some locations, common practice is to mix a four percent solution of sodium fluoride and feed that into the flow of the water system. The amount that is fed depends on the air temperature and on the fluoride levels in the raw water.

Experience has shown that drinking water containing a proper amount of fluoride can reduce tooth decay by 65% in children. Fluoride combines chemically with tooth enamel when permanent teeth are forming. The result, of course, is teeth that are harder, stronger, and more resistant to decay. The USEPA sets the upper limits for fluoride in drinking water supplies based on ambient temperatures, because people drink more water in warmer climates. Fluoride concentrations should be lower in these areas (Spellman, 1998).

11.12 WATER TREATMENT OF ORGANIC AND INORGANIC CONTAMINANTS

Man-made compounds that contain carbon (commonly called SOCs or synthetic organic chemicals) are, from time to time, detected in U.S. water supplies. Some of these are volatile organic compounds (VOCs), such as the solvent trichloroethylene. The problem with VOCs in a water supply used by the public is twofold. They are easily absorbed through the skin, and they volatize into gases, which can then be inhaled by those using showers or baths, or while washing dishes.

How do water supplies become contaminated by organic compounds? Basically, sources of organic contaminants are usually provided by improperly disposed wastes, pesticide use, industrial effluents, and leaking fuel oil tanks (gasoline in particular).

Water supplies may also contain inorganic contaminants consisting mainly of substances occurring naturally in the ground, such as sulfate, fluoride, arsenic, barium, radium, selenium, and radon. Metallic substances from industrial sources can contaminate surface waters. The inorganic ion nitrate (from fertilizers and feedlot runoff in agricultural areas) occurs frequently in groundwater supplies. Another source of inorganic chemical contamination in drinking water supplies is corrosion or deterioration of water supply appurtenances such as plumbing systems, which release metal and nonmetal substances into the water, such as lead, cadmium, zinc, copper, iron, and plumbing cement.

Inorganic contaminants can be treated by corrosion controls and removal techniques. Corrosion controls work to reduce the presence of corrosion by-products (lead, for example) at the consumer's tap. Removal technologies, coagulation/filtration, reverse osmosis (RO), and ion exchange are used to treat source water that is contaminated with metals or radioactive substances.

The following sections discuss processes for removing inorganic and organic dissolved solids from water intended for potable use. Keep in mind that, with some modifications, these same processes may act as tertiary treatment for wastewater.

11.12.1 AERATION

Aeration (air stripping) is a physical treatment process in which air is thoroughly mixed with water—a technique effective for removing dissolved gases and highly volatile odorous compounds. Contact with air and oxygen can improve water quality in a number of ways. For example, when aeration is a first step in processing well water, it may achieve any or all of the following: removal of hydrogen sulfide, reduction of dissolved carbon dioxide, and addition of dissolved oxygen for oxidation of iron and manganese (the oxygen in the air reacts with the iron and manganese to form an insoluble precipitate, rust). One of the most common uses of aeration is for taste and odor control. Sedimentation and filtration are then necessary to clarify the water.

Note: Aeration is rarely effective in processing surface waters, because the odor-producing substances are generally nonvolatile.

Several methods to aerate the water are available. The method selected depends primarily on the type and concentration of material to be removed from the water and on the available pressure. Aeration in water treatment can be accomplished using spray nozzles, cascade systems, multiple-tray aerators, diffused-air aerators, and mechanical aerators.

11.12.2 OXIDATION

Oxidation is a reaction in which a substance loses electrons, thus increasing its charge. A substance that oxidizes another is called an *oxidizing agent* or *oxidizer*. In water treatment, oxidation is used to remove or destroy undesirable tastes or odors, to aid in removal of iron and manganese, and to help improve clarification and color removal in source water.

Note: Atmospheric oxygen through aeration can be used to oxidize the organic substances responsible for undesirable tastes and odors, but the process is usually too slow to be of value. However, if dissolved gases such as hydrogen sulfide are the cause of taste and odor problems, aeration will effectively remove them through oxidation and stripping.

Chlorine dioxide, potassium permanganate, and ozone are strong oxidants capable of destroying many odorous compounds. Because they do not produce THMs, these chemicals are favored over heavy chlorination.

11.12.3 ADSORPTION

When we speak of *adsorption*, we discuss primarily a surface phenomenon: the adsorption that results when one substance attaches itself to the surface of another. The two most common adsorptive media used in water treatment are activated carbon and activated alumina. These adsorptive materials are generally most effective for taste and odor control, and for removal of organic pollutants. However, the most important applications of adsorption in water treatment are the removal of arsenic and organic pollutants.

Adsorption of organic materials using *activated carbon* has been a common practice in water treatment for many years. Activated carbon is manufactured from carbonaceous material such as wood, coal, or petroleum residues. A char is made by burning the material in the absence of air, which is then oxidized at higher temperatures to create a very porous structure. This "activation" step provides irregular channels and pores in the solid mass, resulting in a very large surface-area-per-mass ratio. This large surface area gives activated carbon its effectiveness as an adsorbing agent. The larger the surface area of an adsorber, the greater its power. Each activated carbon contains a huge number of pores and crevices into which organic molecules enter and are adsorbed onto the activated carbon surface.

Activated carbon has a particularly strong attraction for organic molecules such as the aromatic solvents benzene, toluene, nitrobenzene; the chlorinate aromatics PCBs, chlorobenzenes, chloronaphthalene; phenol and chlorophenols; the polynuclear aromatics acenaphthene and benzopyrenes; pesticides and herbicides; chlorinated aliphatics such as carbon tetrachloride, chloroalkyl ethers; and high molecular weight hydrocarbons such as dyes, gasoline, amines, and humics.

Two forms of activated carbon are used in water treatment: powdered and granular. Powdered activated carbon is often used for taste and odor control. Its effectiveness depends on the source of the undesirable tastes and odors. It is also effective in removing the organic precursors that react with chlorine to form harmful THM compounds after disinfection.

Powdered activated carbon is a finely ground, insoluble black powder that can be added at any point in the treatment process ahead of the filters. It is fed to water either as a dry powder or as a wet slurry. Although adsorption is nearly instantaneous, a contact time of 15 minutes or more is desirable before sedimentation or filtration.

Activated carbon media must periodically be replaced with new or regenerated activated carbon. Replacement cycles can vary from one to three years for taste and odor treatment to as little as four or five weeks for removal of organics. The activated carbon regeneration process involves: (1) removing the spent carbon as a slurry; (2) dewatering the slurry; (3) feeding the carbon into a special furnace where regeneration occurs (i.e., the organics are driven from the carbon surface by heat); and (4) returning it to use.

Activated alumina (a highly porous and granular form of aluminum oxide) is also an adsorptive medium used in water treatment, primarily to remove arsenic and excess fluoride ions. In use, water is percolated through a column of alumina media. A combination of adsorption and ion exchange performs the actual removal of arsenic and fluoride ions.

Like the regeneration process used to restore used activated carbon to full potency, activated alumina also requires periodic regeneration, accomplished by passing a caustic soda solution through the media. Excess caustic soda is neutralized by rinsing the activated alumina with an acid. Disposal of these wash waters, laden with toxic arsenic and fluoride ions, must be done in accordance with applicable laws.

Note: Powdered activated carbon is much more difficult to regenerate than granular activated carbon. Granular activated carbon is sometimes used in the filter bed itself, combining both filtration and adsorption in one treatment unit. The major problem associated with granular activated carbon systems is plugging of the bed by suspended solids in the water.

11.12.4 DEMINERALIZATION

Demineralization refers to the removal of dissolved solids (inorganic mineral substances) from water. Dissolved solids contain both cations and anions and therefore require two types of ion exchange resins. Cation exchange resins used for demineralization purposes have hydrogen exchange sites and are divided into strong acid and weak acid classes. The anion exchange resins commonly used contain hydroxide ions and are divided into strong and weak base classes.

Demineralization is commonly used in industry in waste treatment for removal of arsenic, barium, cadmium, chromium, fluoride, sulfate, and zinc. Some general advantages of using ion exchange to remove these contaminants are the low capital investment required and the mechanical simplicity of the process. In addition, the ion exchange process can be used to recover valuable chemicals for reuse, or harmful ones for disposal. For example, it is often used to recover chromic acid from metal finishing waste for reuse in chrome-plating baths. It also has some application in the removal of radioactivity. The major disadvantages are the high chemical requirements needed to regenerate the resins and to dispose of chemical wastes from the regeneration process. These factors make ion exchange more suitable for small systems than for large ones.

11.12.5 MEMBRANE PROCESSES

Membrane processes used in water treatment are primarily demineralization processes. Demineralization of water can be accomplished using thin, microporous membranes. Electrodialysis and reverse osmosis are the most common membrane processes.

During osmosis, two solutions containing different concentrations of minerals are separated by a semipermeable membrane. Water tends to migrate through the membrane from the side of the more dilute solution to the side of the more concentrated solution. This is *osmosis*, and it continues until the buildup of hydrostatic pressure on the more concentrated solution is sufficient to stop the net flow.

In *reverse osmosis*, the flow of water through the semipermeable membrane is reversed by applying external pressure to offset the hydrostatic pressure. This results in a concentration of minerals on one side of the membrane and pure water on the other side. Reverse osmosis can treat for a wide variety of health and aesthetic contaminants in water. Effectively designed, reverse osmosis equipment can treat aesthetic contaminants that cause unpleasant taste, color, and odor problems, like a salty or soda taste caused by chlorides or sulfates. Reverse osmosis can also be effective for treating arsenic, asbestos, atrazine, fluoride, lead, mercury, nitrate, and radium. When used with appropriate carbon prefiltering, additional treatment can also be provided for such "volatile" contaminants as benzene, trichloroethylene, trihalomethanes, and radon. Some reverse osmosis equipment is also capable of treating for *Cryptosporidium*. Reverse osmosis can be expected to play a major role in water treatment for years to come.

Reverse osmosis (also called ultrafiltration) is the most common process for reducing the salinity of brackish groundwater. In operation, a semipermeable membrane (the most essential element in the reverse osmosis method of demineralization) separates salty water of two different concentrations. Concentrations have a natural tendency to become equalized by a flow of water from the dilute side to the concentrated side (osmosis). But high pressure applied to the high concentration side of the membrane can reverse this direction of flow. Freshwater diffuses through the membrane, leaving a more concentrated salt solution behind. The performance of reverse osmosis units is highly dependent on a number of water quality parameters. Suspended solids, dissolved organics, hydrogen sulfide, iron, and strong oxidizing agents (chlorine, ozone, and permanganate) are harmful to membranes.

Electrodialysis is the demineralization of water using the principles of osmosis, but it uses ion-selective membranes and an electric field to separate anions and cations in solution. In the past, electrodialysis was most often used for purifying brackish water, but it is now finding a role in industrial waste treatment as well. For example, metals salts from plating rinses are sometimes removed in this way.

11.13 SUMMARY

These processes, from preliminary screening to filtration, disinfection, and the advanced methods needed for specialized water problems, are in place to serve one primary, essential purpose: to supply the consumer with safe potable water. The consumer may put this water to a wide variety of uses, from drinking to watering the lawn, but those uses are in many ways beside the point: safe potable water is essential for human life.

11.14 REFERENCES

AWWA. *Water Treatment: Principles and Practices of Water Supply Operations.* Denver, CO: American Water Works Association, 1995.

De Zuane, J., *Handbook of Drinking Water Quality.* 2nd ed. New York: John Wiley and Sons, 1997.

Fox, J. C. Stakeholder Collaboration Key to New Safe Drinking Water Rules, *Water Environment & Technology,* February 1999.

Hammer, M. J. and Hammer, M. J., Jr. *Water and Wastewater Technology,* 3rd ed. Englewood Cliffs, NJ: Prentice Hall, 1996.

Larsen, T. J., *Water Treatment Plant Design.* New York: Cox Publishing, 1990.

Masters, G. M., *Introduction to Environmental Engineering and Science.* Englewood Cliffs, NJ: Prentice Hall, 1991.

McGhee, T. J., *Water Supply and Sewerage.* 6th ed. New York: McGraw-Hill, Inc., 1991.

Pankratz, T. M., *Screening Equipment Handbook: For Industrial and Municipal Water and Wastewater Treatment,* 2nd ed. Lancaster, PA: Technomic Publishing Company, Inc., 1995.

Peavy, H. S. et al., *Environmental Engineering.* New York: McGraw-Hill, Inc., 1985.

Spellman, F. R., *The Science of Water: Concepts and Applications.* Lancaster, PA: Technomic Publishing Company, Inc., 1998.

Spellman, F. R., *Disinfection Alternatives.* Lancaster, PA: Technomic Publishing Company, Inc., 1999.

USA TODAY, October 1998.

Viessman, W. and Hammer, M. J., *Water Supply and Pollution Control,* 6th ed. Menlo Park, California: Addison-Wesley, 1998.

Afterword

A S our world population grows and our available safe water supply must meet increasingly steep demands, we foresee widespread growing concern over water quality.

Human nature dictates that we do not value what we have until it is at risk, no matter how essential to us it is. We are utterly dependent on the presence and quality of our drinking water; and after decades of benign neglect (a contradiction in terms if we ever heard one) of our water supply and the infrastructure essential to producing it, we are becoming alert to the fact that our essential potable water supplies are at risk in several ways, from overuse to pollution to process and equipment failure.

Many people are already expressing their concern in a number of ways. They purchase and install home filtration systems. They check for elevated lead levels in the public water supply. They have their well water tested. Some public water system consumers are already making alternative choices concerning their water supply. They buy bottled drinking water. For a wide variety of reasons, they've decided not to drink the water most readily available to them. A few towns in the U.S. cannot at present provide safely potable public water. For some consumers it is a matter of politics, local public policy, philosophy, and aesthetics. Some consumers detect a chemical taste that displeases them in their "city water." They may take issue with fluoridation. They fear microbiological contamination. They've been reading about chlorine by-products. They like the assurance of that certification on the label. They like the implied status of drinking bottled water.

Some of these reasons would seem trivial to the person with no public water supply available, who hauls water from a well in the village square, or who carries water straight from the local surface water sources, wherever they may be.

Our public water treatment systems in general are excellent, though not perfect. In many places, equipment is aging, and population growth demands expansion. New processes and technologies should replace old in many locations. However, while we recognize that we need improvements, changes, and new technologies, we also recognize how fortunate we are. Our basic infrastructure is in place and working. Serious problems with water quality and supply are the exception, not the rule. Here in the U.S., regulation demands a high level of quality for our water and protects our supplies. To state it simply, the problems we are attempting to solve aren't the fundamental ones: the scrabbling for the funds and scrambling for the technologies essential to developing public water treatment in the first place. The new amendments to the Safe Drinking Water Act will help to solve some infrastructure and expansion problems, to upgrade many small systems, and to handle details of water safety and quality that those with no access to safe public water would be glad to have as problems instead.

That's a point to consider when you turn on the tap and fill your glass, when you brush your teeth, run the dishwasher, step into the shower, wash your car, or water your lawn. Our water is an essential, limited resource—but those in plenty may not recognize it for the treasure it is.

Index